Signal Processing: Fundamentals and Advanced Techniques

Signal Processing: Fundamentals and Advanced Techniques

Jhonny Jones

Willford Press
www.willfordpress.com

Published by Willford Press,
118-35 Queens Blvd., Suite 400,
Forest Hills, NY 11375, USA

ISBN: 978-1-64728-501-2

Cataloging-in-Publication Data

Signal processing : fundamentals and advanced techniques / Jhonny Jones.
p. cm.
Includes bibliographical references and index.
ISBN 978-1-64728-501-2
1. Signal processing. 2. Signal theory (Telecommunication). 3. Signal processing--Equipment and supplies. I. Jones, Jhonny.
TK5102.9 .S54 2023
621.382 2--dc23

For information on all Willford Press publications
visit our website at www.willfordpress.com

Willford Press

Contents

Preface

It is often said that books are a boon to mankind. They document every progress and pass on the knowledge from one generation to the other. They play a crucial role in our lives. Thus I was both excited and nervous while editing this book. I was pleased by the thought of being able to make a mark but I was also nervous to do it right because the future of students depends upon it. Hence, I took a few months to research further into the discipline, revise my knowledge and also explore some more aspects. Post this process, I begun with the editing of this book.

Signal processing is a subfield of electrical engineering involved in analyzing, modifying and synthesizing signals such as sound, images and scientific measurements. The techniques of signal processing are used to improve transmission, storage efficiency, and subjective quality. They can also emphasize or detect components of interest in a measured signal. There are various categories of signal processing such as analog, continuous time, digital, discrete time, non-linear and statistical signal processing. Analog signal processing refers to the signal processing which is conducted on continuous analog signals through analog means. The techniques which are used to improve the accuracy and quality of digital communication is known as digital signal processing. Some of the prominent applications of signal processing are economic modeling and analysis, weather forecasting, oceanography, medical imagery (such as CAT and MRI), oil explorations, seismic studies, and data analysis. The objective of this book is to give a general view of signal processing with an emphasis on advanced techniques. It is an essential guide for both students and professionals.

I thank my publisher with all my heart for considering me worthy of this unparalleled opportunity and for showing unwavering faith in my skills. I would also like to thank the editorial team who worked closely with me at every step and contributed immensely towards the successful completion of this book. Last but not the least, I wish to thank my friends and colleagues for their support.

Jhonny Jones

1

Introduction to DSP System, Iteration Bound, Pipelining and Parallel Processing

1.1 Introduction to Digital Signal Processing System

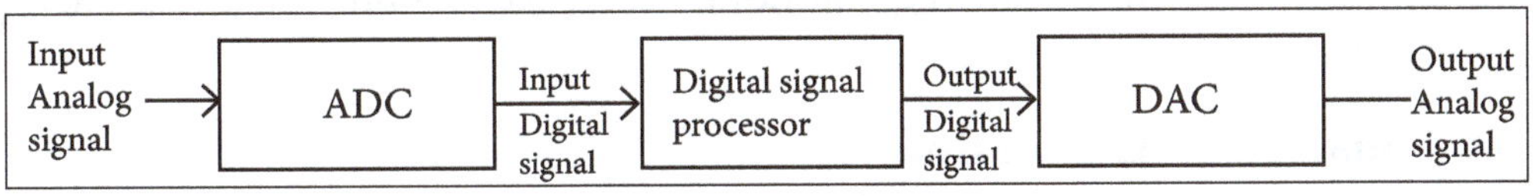

Block diagram of DSP.

ADC

It converts analog input to digital input.

Digital Signal Processor

It performs amplification, attenuation, filtering, spectral analysis, feature extraction etc operations on digital data.

DAC

Some of the processed signals are required back in the analog form. So we use DAC to convert digital output to its analog equivalent.

Advantages

- Greater accuracy.
- Cheaper.
- Ease of data storage.
- Implementation of sophisticated algorithms.
- Flexibility in configuration.

- Applicability of VLF signals.
- Time sharing.

Application of DSP

DSP for Speech and Voice

Speech recognition, Voice mail, Speech coding, Speaker verification, Speech enhancement, Speech synthesis, Text to speech.

DSP for Telecommunication

FAX, Cellular phones, Digital speech interpolation, Video conferencing, Spread spectrum communication, Packet switching, echo cancelling, Digital EPABXs, AOPCM transponders, Channel multiplexing, Modems adaptive equalizers, Data encryption and line repeaters.

DSP for Consumers Application

Digital Audio, Video, Television, Music system, Music synthesizes, Toys.

DSP for Graphics and Imaging

3D and 2D visualization, animation, pattern recognition, Image transmission and compression, Image enhancement, Robot Vision, Satellite imaging for multipurpose application.

1.1.1 Representation of DSP Algorithms

DSP algorithms can be represented in the following ways:

- Behavioral description.
- Graphical representation.

Behavioral Description

- Applicative language: Represents a set of equations satisfied by the variables.
- Perspective language: Explicitly specify the order of assignment. Example, C and other HLLs.
- Descriptive language: Represents the structure of a DSP system. Example, VHDL and Verilog.

Graphical Representation

- Block diagram

- Signal-Flow Graph(SFG)
- Data-Flow Graph (DFG)
- Dependence Graph (DG)

Block Diagram

The block diagram of a system is a pictorial representation of the functions performed by each component of the system and shows the flow of signals. The basic elements of block diagram are as follows:

- Block
- Branch point
- Summing point

Block: The block is the symbolic representation of transfer function of that element. The complete control system can be represented with a required number of interconnection of such blocks.

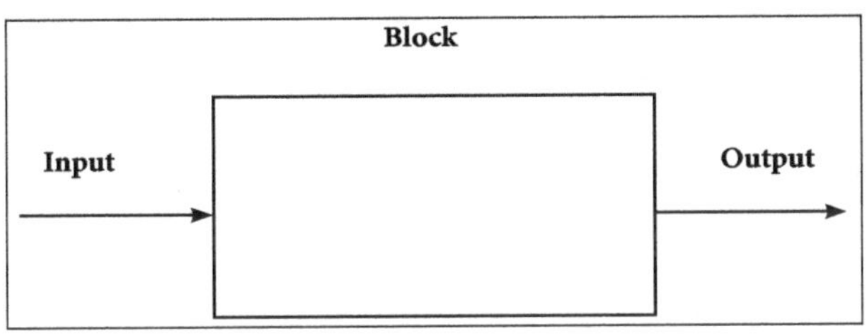

Block

Summing Point: Different input signals are applied to same block. Here, resultant input signal is the summation of all input signals applied. Summation of input signals is represented by a point called summing point which is shown in the below figure by crossed circle.

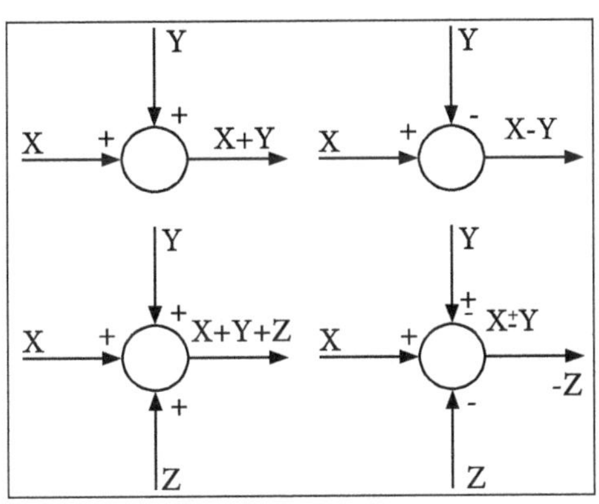

Summing point

Block Diagram Rules

1. Cascaded blocks:

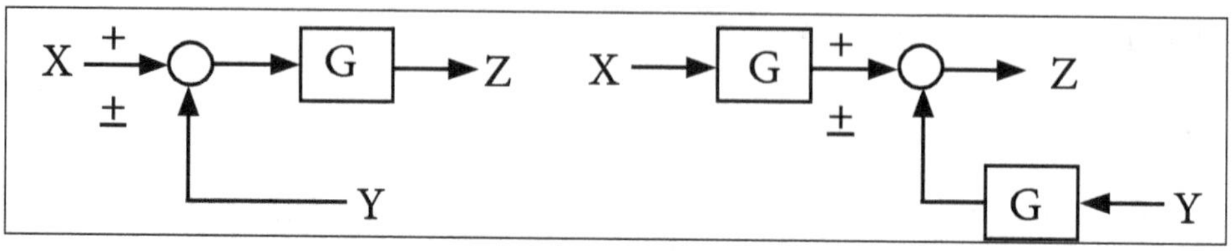

2. Moving a summer beyond the block:

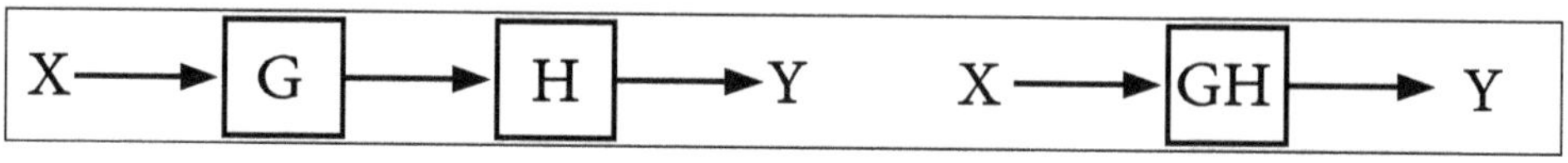

3. Moving a summer ahead of block:

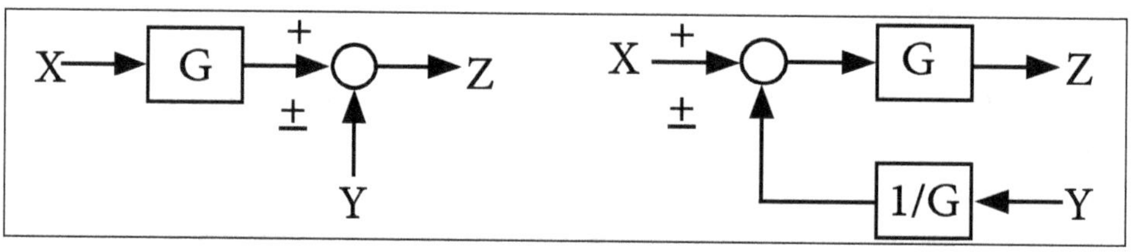

4. Moving a pick-off ahead of block:

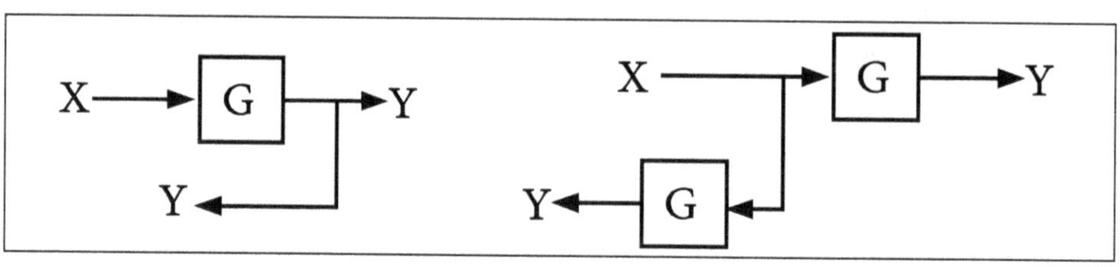

5. Moving a pick-off behind a block:

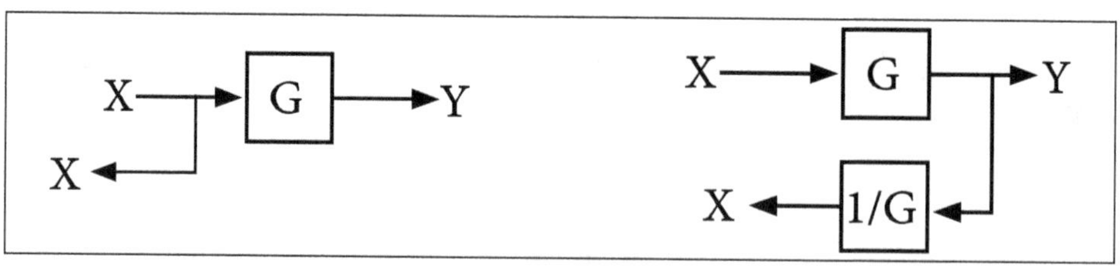

6. Eliminating a feedback loop:

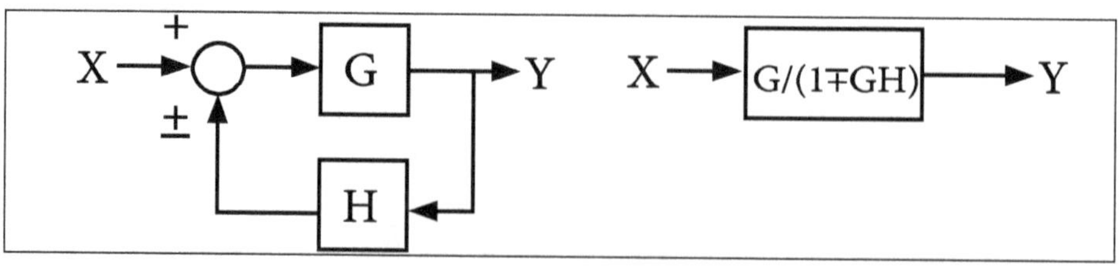

7. Cascaded Subsystems:

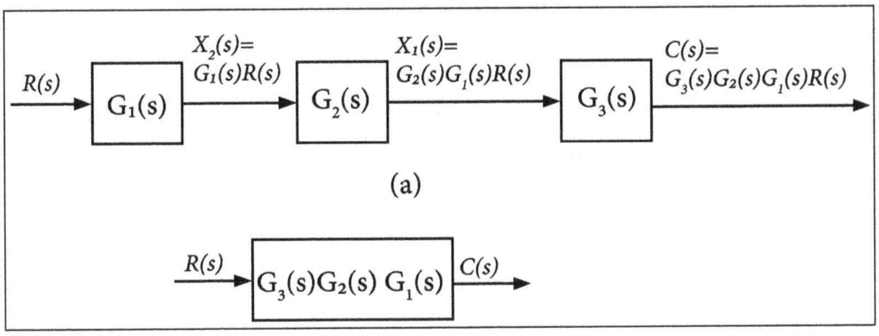

8. Parallel Subsystems:

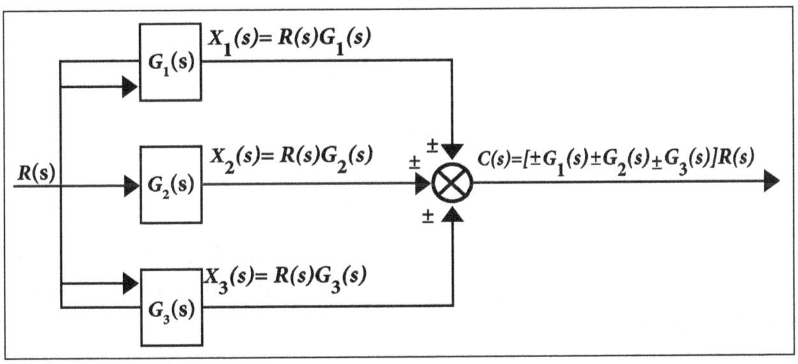

Signal Flow Graph

A signal flow graph is the diagram that represents a set of simultaneous linear algebraic equations. By taking Laplace transform, the time domain differential equations can be transferred to a set of algebraic equation in s-domain. The signal-flow graph consists of a network in which nodes are connected by directed branches. The depicts the flow of signals from one point of a system to another and gives the relationships among the signals.

Properties of Signal Flow Graph

- Signal flow applies only to linear systems.
- The equations based on which a signal flow graph is drawn must be algebraic equations in the form of effects as a function of causes.
- Signals travel along the direction described by the arrows of the branches.

Procedure for Determining Transfer Function using Signal Flow Graph

- Determine the number of forward paths and determine the forward path gain correspondingly.

- From the given graph, find the possible individual loop and corresponding gains for them. Similarly find the number of two, three etc., non-touching loops and its gain.
- Calculate the value of Δ and Δ_K. Substitute all the gains in the Mason's gain formula to determine the transfer function.

Mason's Gain Formula

$$T = \frac{1}{\Delta}\sum_K P_K \Delta_K$$

Where,

$T = T(s)$ = Transfer function of the system.

K = Number of forward paths in the signal flow graph.

P_K = Forward path gain of Kth forward path.

$\Delta = 1 -$ [Sum of individual loop gains] + [Sum of gain product of all possible combination of two non-touching loops] - [Sum of gain product of all possible combination of three non-touching loops] + ...

$\Delta_K = \Delta$ For that part of the graph which is not touching the kth forward path.

Table: Comparison between block diagram and signal flow graph methods:

S. No.	Block Diagram	Signal Flow Graph
1.	Each element is represented by a block.	Each variable is represented by a separate node.
2.	Transfer function of the element is given inside the block.	Transfer function is shown along the branches connecting the nodes.
3.	Summing points and take off points are separate.	Any node can have number of incoming and outgoing branches.
4.	Feedback path is present.	Feedback loops are considered for the analysis.
5.	Self-loop concept is absent.	Self-loop can exist.
6.	Applicable to known invariant systems.	Applicable to linear time invariant systems.

Problems

1. Let us calculate the closed-loop transfer function C/R using block diagram reduction technique whose block diagram is shown below:

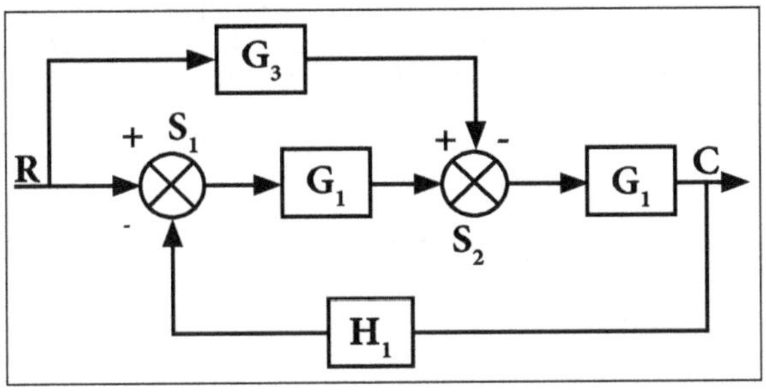

Solution:

Given:

1. Shift the summing point S_1 after block G_1:

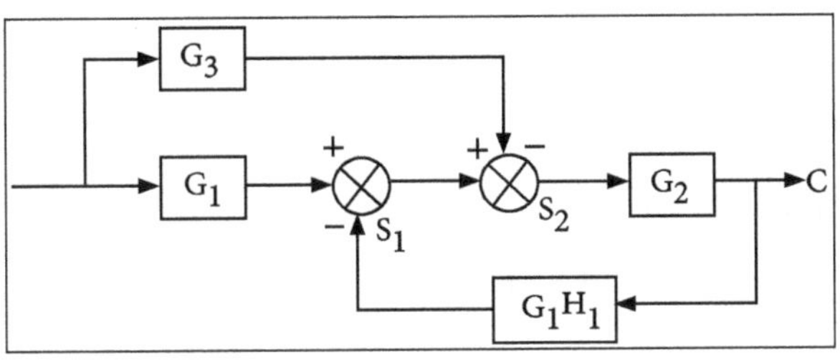

2. Interchanging S_1 and S_2:

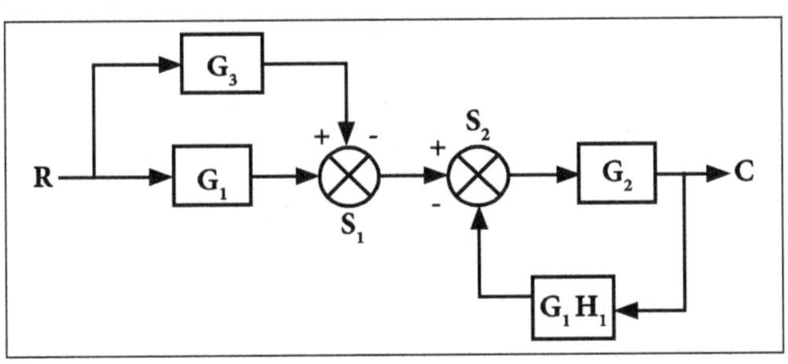

3. Eliminate S_1 point and S_2:

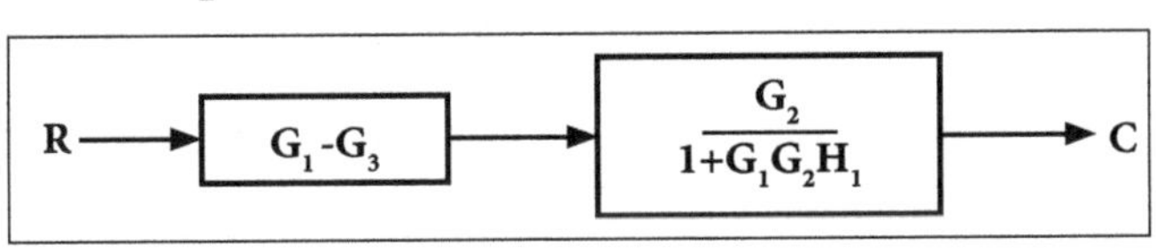

$$\therefore \text{Transfer function, } \frac{C(s)}{R(s)} = \frac{(G_1 - G_3)G_2}{1 + G_1G_2H_1}$$

2. Using block diagram reduction method, let us determine the output of the below system:

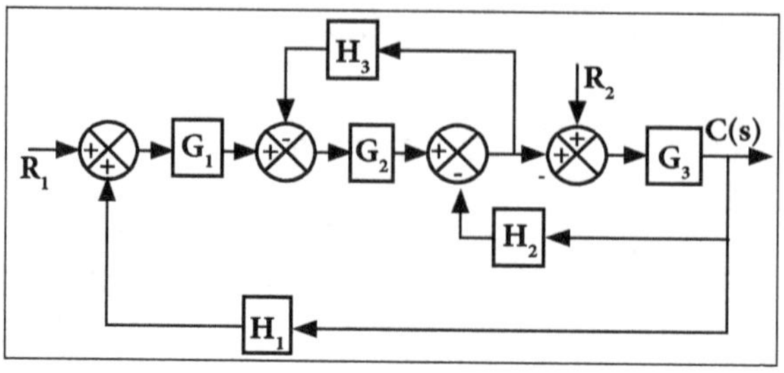

Solution:

Given:

Transfer function of the above block diagram is given by,

The Transfer function becomes $= \frac{C}{R_1} + \frac{C}{R_2}$

$$\frac{C}{R_1} = \frac{G_1G_2G_3}{1 + H_1G_1G_2G_3 + G_3H_2 + G_2H_3}$$

$$\frac{C}{R_2} = \frac{G_3(1 + G_2H_3)}{1 + H_1G_1G_2G_3 + G_3H_2 + G_2H_3}$$

Step 1: Shift the takeoff point before the G_4 block to a point after the block.

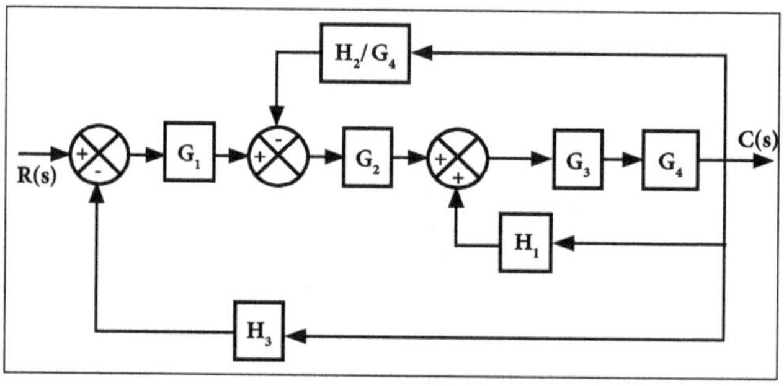

Step 2: Combine cascade blocks of G_3, G_4 and eliminate the closed loops formed by G_3 G_4 and H_1.

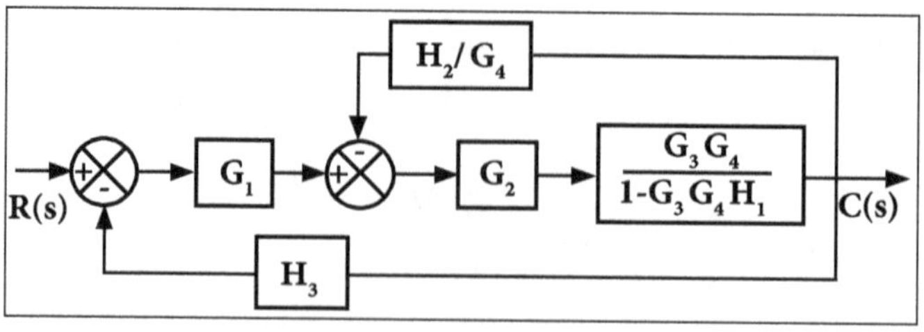

Step 3: Combine the blocks in cascade with G_2 and $\frac{G_3G_4}{1-G_3G_4H_1}$ then, eliminate the closed loop formed by them with H_2/G_4.

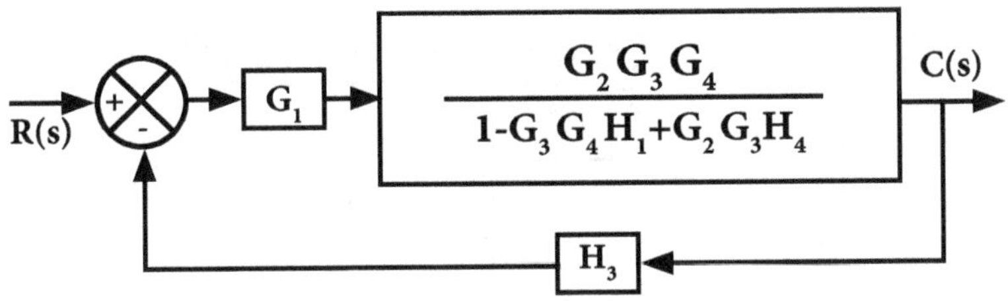

Step 4: Combine the blocks in cased with G_1 and $\frac{G_2G_3G_4}{1-G_3G_4H_1+G_2G_3H_2}$.

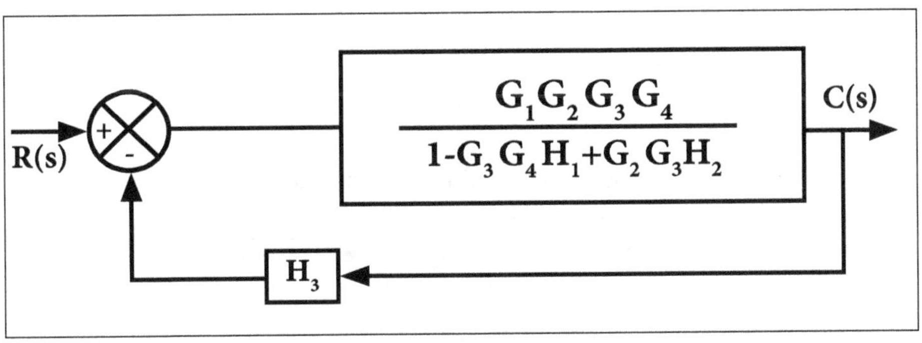

Step 5: Eliminate the feedback loop formed by H_3. The Transfer function of the given block diagram is given by,

$$\frac{C}{R}=\frac{G_1G_2G_3G_4}{1-G_3G_4H_1+G_2G_3H_2+G_1G_2G_3G_4H_3}$$

3. Let us calculate C(s)/R(s) for the below figure by reducing the block.

Solution:

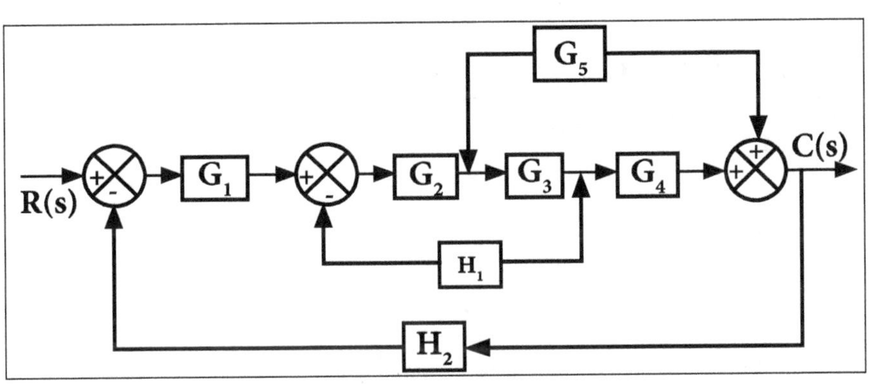

Given:

Step 1: Shift the takeoff point after the block G_3 to before G_3.

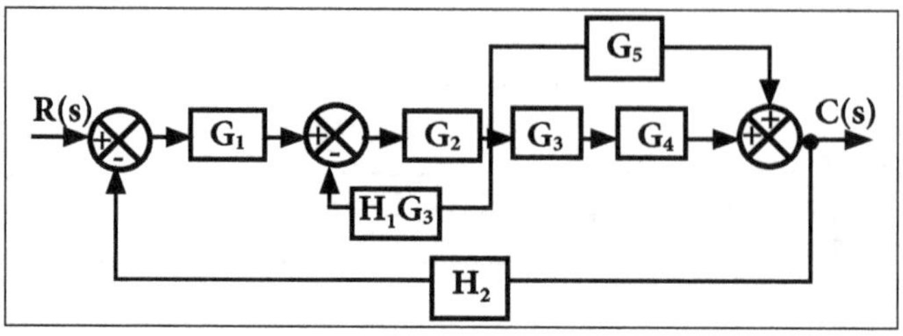

Step 2: Combine the cascade blocks G_3, G_4 and eliminates the parallel blocks.

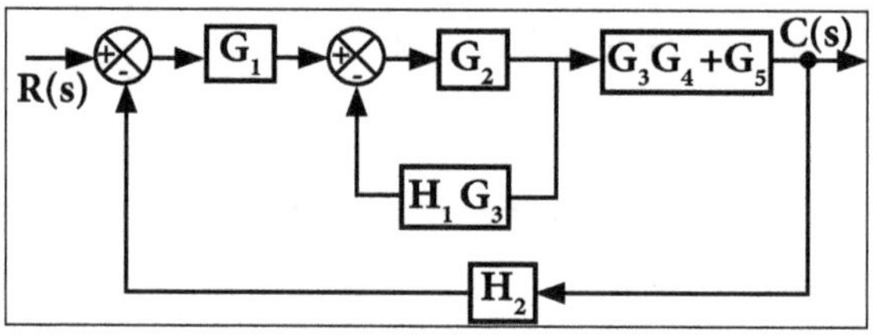

Step 3: Eliminate the feedback loop formed by G_2 and H_1G_3.

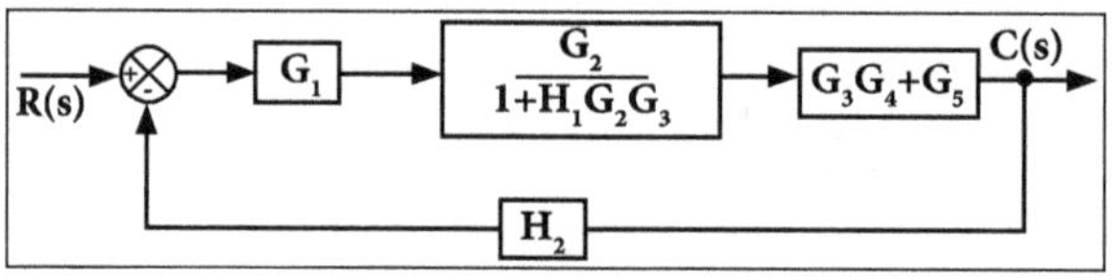

Step 4: Combine the cascade blocks.

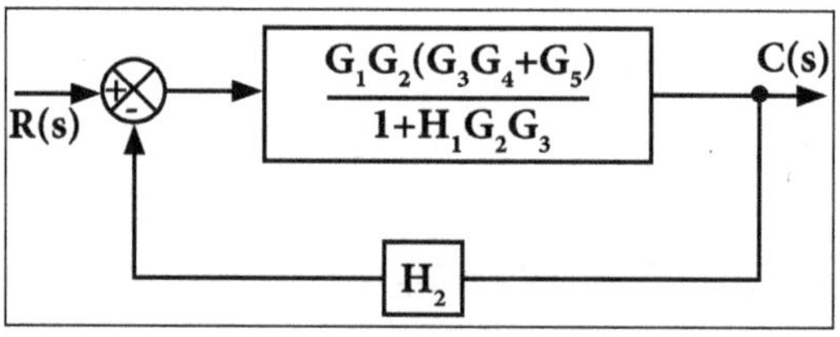

Step 5: Eliminate the feedback loop.

$$\frac{C(s)}{R(s)} = \frac{G_1G_2(G_3G_4+G_5)/1+H_1G_2G_3}{1+H_2\times\dfrac{G_1G_2(G_3G_4+G_5)}{1+H_1G_2G_3}}$$

The transfer function becomes,

$$\frac{C(s)}{R(s)} = \frac{G_1G_2(G_3G_4+G_5)}{1+H_1G_2G_3+\left[G_1G_2(G_3G_4+G_5)\right]H_2}$$

4. By Reducing the block diagram for below figure, let us obtain its closed loop transfer function C(S)/R(S).

Solution:

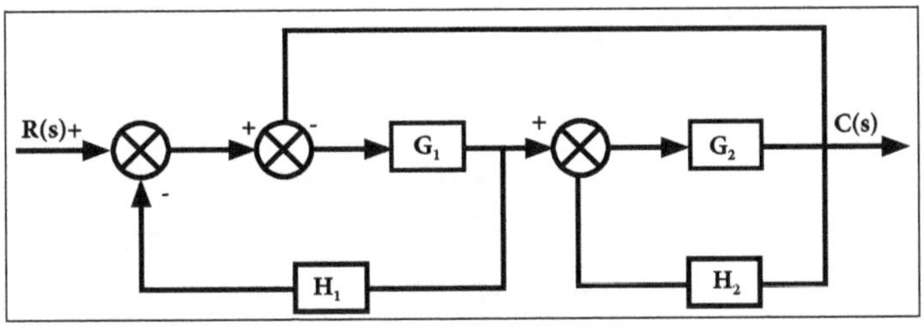

Given:

Step 1: Eliminate the closed loop formed by G_2 and H_2, then the circuit diagram becomes,

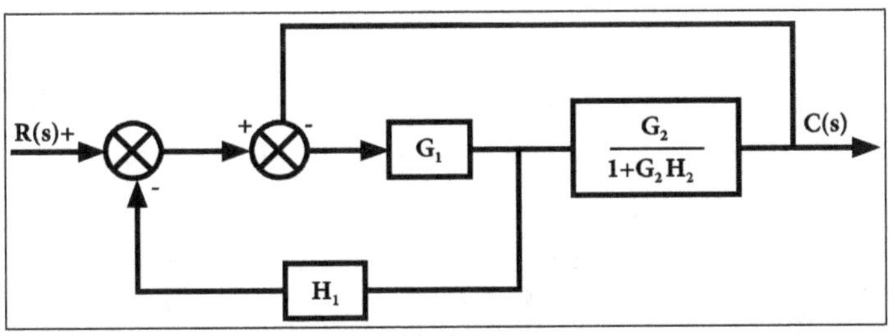

Step 2: Shift the takeoff point H_1 before the block $\frac{G_2}{1+G_2H_2}$ to a point after the block,

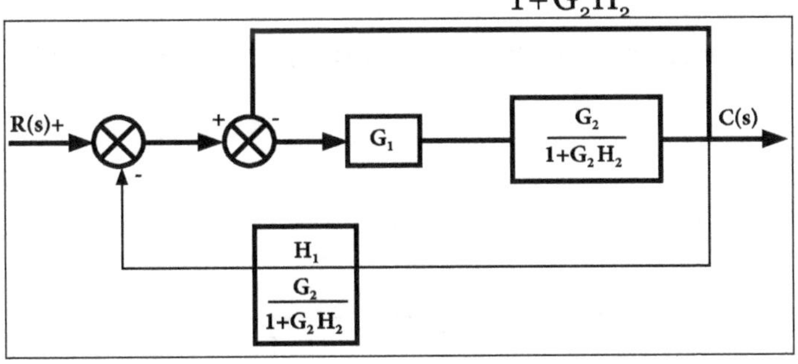

Step 3: Combine the cascade blocks G_1 and $\frac{G_2}{1+G_2H_2}$, then,

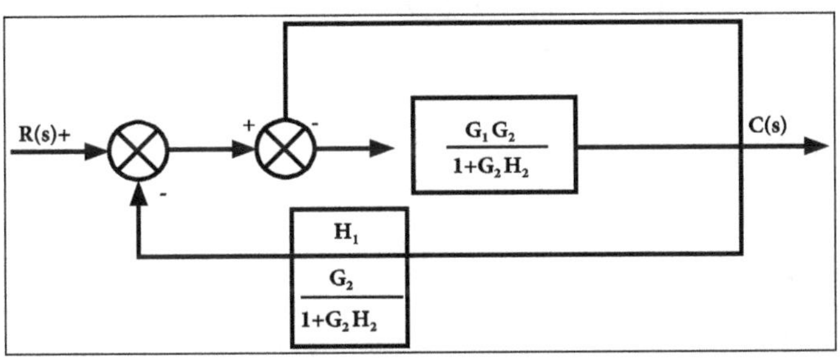

Step 4: Eliminate the Closed loop formed by $\frac{G_1G_2}{1+G_2H_2}$, then,

$$\frac{\frac{G_1G_2}{1+G_2H_2}}{1+\frac{G_1G_2}{1+G_2H_2}}=\frac{G_1G_2}{1+G_2H_2+G_1G_2}$$

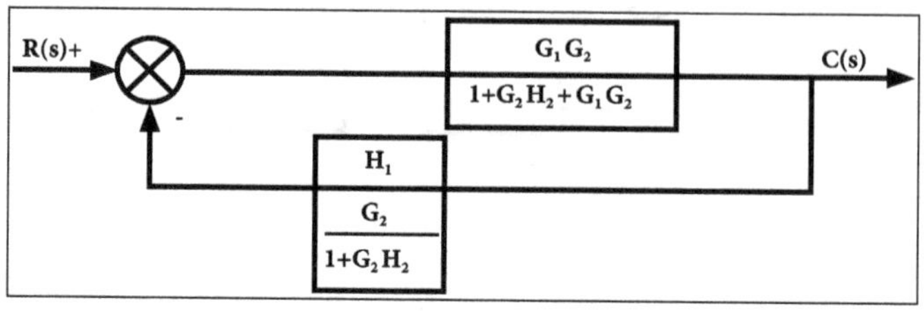

Step 5: Eliminate the closed loop formed by $\frac{G_1G_2}{1+G_2H_2+G_1G_2}$ and $\frac{H_1+G_2H_1H_2}{G_2}$.

$$\frac{C(s)}{R(s)}=\frac{\frac{G_1G_2}{1+G_2H_2+G_1G_2}}{1+\frac{G_1G_2}{1+G_2H_2+G_1G_2}\times\frac{H_1+G_2H_1H_2}{G_2}}=\frac{G_1G_2}{1+G_2H_2+G_1G_2+G_1H_1+G_1G_2H_1H_2}$$

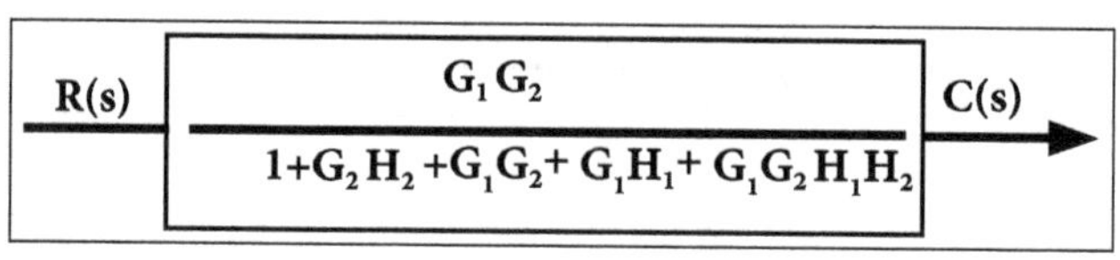

5. Let us derive the overall transfer function of the following block diagram using signal flow graph method.

Solution:

Given:

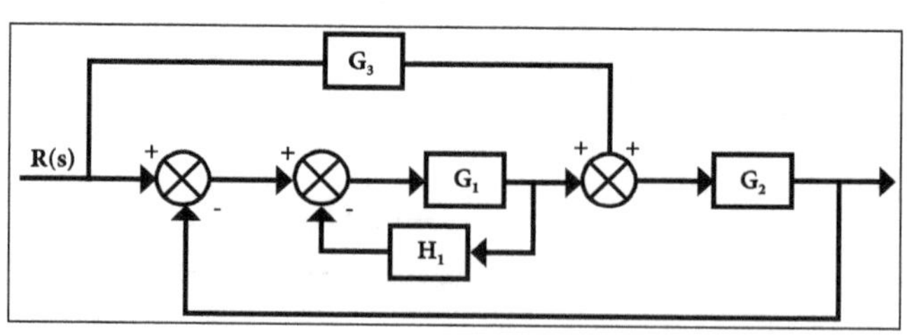

To find:

Transfer function:

$$\frac{C(s)}{R(s)}=\frac{G_1G_2+G_3G_2}{1+G_1G_2+G_3G_2+G_1H_1}$$

6. Let us calculate the transfer function of the transistor's hybrid model below using signal flow graph.

Solution:

Given:

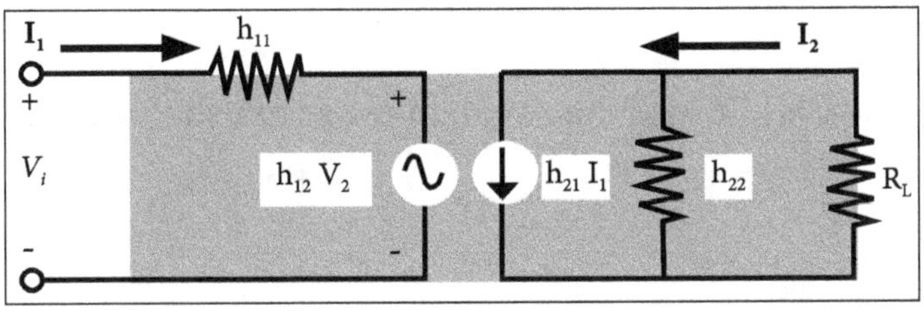

We know that the hybrid parameter equation for a transistor is given by,

$$V_1 = h_{11}I_1 + h_{12}V_2$$

$$I_2 = h_{21}I_1 + h_{22}V_2$$

Output voltage, $V_2 = I_2R_L$

The above model can be represented in the signal flow graph as,

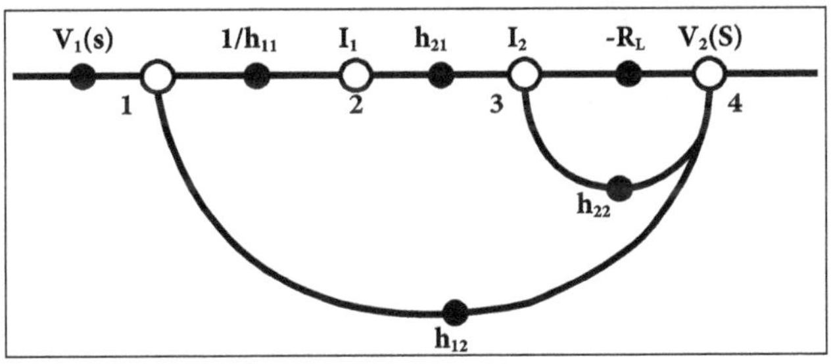

From the Signal Flow Graph:

There is only one forward path, $P_1 = \dfrac{-h_{21}R_L}{h_{11}}$

There are two loops such as:

Loop formed by node 1, $L_1 = \dfrac{h_{12}h_{21}R_L}{h_{11}}$

Loop formed by node 3, $L_2 = h_{22}R_L$

All the loops touching the forward path, path factor $\Delta_1 = 1$

$$\Delta = 1-(L_1+L_2) = 1-h_{22}R_L - \frac{h_{12}h_{21}R_L}{h_{11}}$$

By using Mason's gain formula, the transfer function of transistor hybrid model is given by,

$$T.F = \frac{P_1\Delta_1}{\Delta} = \frac{-\dfrac{h_{21}R_L}{h_{11}}}{1-h_{22}R_L-\dfrac{h_{12}h_{21}R_L}{h_{11}}} = \frac{-h_{21}R_L}{h_{11}-h_{11}h_{22}R_L-h_{12}h_{21}R_L}$$

7. Let us calculate the C(s)/R(s) for the signal flow graph shown below.

Solution:

Given:

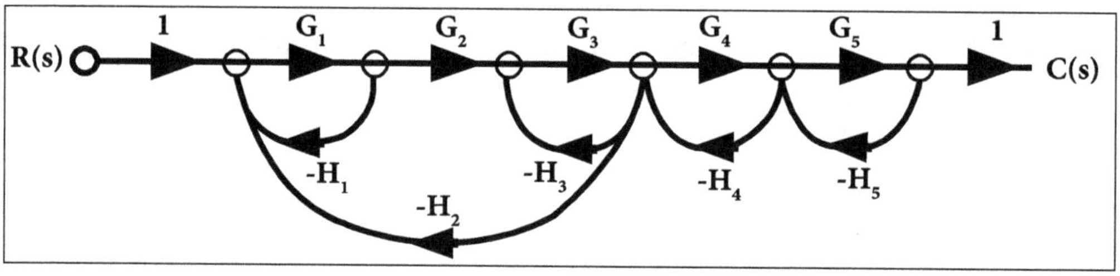

Step 1: The number of forward path k = 1.

Forward path gain $P_1 = G_1G_2G_3G_4G_5$

Step 2: Number of individual loops:

$$L_1 = -G_1H_1$$

$$L_2 = -G_1G_2G_3H_2$$

$$L_3 = -G_3H_3$$

$$L_4 = -G_4H_4$$

$$L_5 = -G_5H_5$$

Step 3: Number of two non-touching loops at a time:

$$L_1L_3 = G_1H_1\,G_3H_3$$

$$L_1L_4 = G_1H_1\,G_4H_4$$

$$L_1L_5 = G_1H_1\,G_5H_5$$

$$L_2L_4 = G_1G_2G_3H_2G_4H_4$$

$$L_2L_5 = G_1G_2G_3H_2G_5H_5$$

$$L_3L_5 = G_3H_3\ G_5H_5$$

Step 4: There are only one, three non-touching loops at a time.

$$L_1\ L_3L_5 = -G_1H_1\ G_3H_3G_5H_5$$

Step 5: By using mason's gain formula, the transfer function becomes,

$$\frac{C(s)}{R(s)} = \frac{\sum P_K \Delta_K}{\Delta}$$

$$= \frac{P_1\Delta_1}{1-(L_1+L_2+L_3+L_4+L_5)+(L_1L_3+L_1L_4+L_1L_5+L_2L_4+L_2L_5+L_3L_5)-(L_1L_3L_5)}$$

8. Let us obtain the transfer function of the below figure using Mason's gain formula.

Solution:

Given:

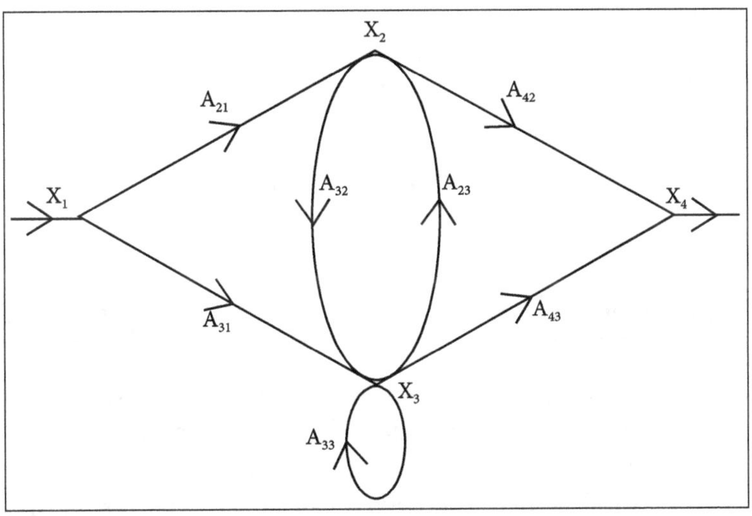

Formula to be used:

Mason's gain formula,

$$T = \frac{1}{\Delta}\sum_K P_K \Delta_K$$

$$\frac{C(s)}{R(s)} = \frac{\sum_1 P_k\ \Delta_k}{\Delta} = \frac{P_1\Delta_1 + P_2\Delta_2 + P_3\Delta_3 + P_4\Delta_4}{1-[L_1+L_2]}$$

$$\frac{C(s)}{R(s)} = \frac{A_{21}A_{42}(1-A_{33})+A_{31}A_{43}+A_{21}A_{32}A_{43}+A_{31}A_{23}A_{42}}{1-(A_{23}A_{32}+A_{33})}$$

9. Let us construct the equivalent signal flow graph and C/R using Mason's gain formula for the below block diagram.

Solution:

Given:

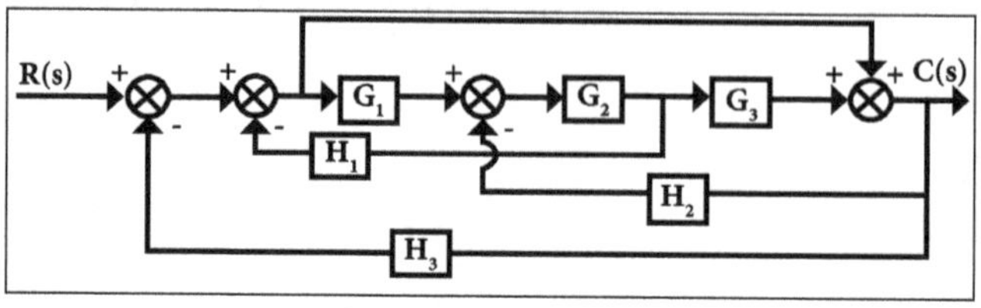

Formula to be used:

Mason's gain formula,

$$T = \frac{1}{\Delta}\sum_{K} P_K \Delta_K$$

$$\frac{C(s)}{R(s)} = \frac{\sum_{k} P_k \Delta_k}{\Delta} = \frac{P_1\Delta_1 + P_2\Delta_2}{1-\left[L_1 + L_2 + L_3 + L_4 + L_5\right]}$$

$$\frac{C(s)}{R(s)} = \frac{1+G_1G_2G_3}{1+G_1G_2G_3H_3 + H_3 + G_1G_2H_1 + G_2G_3H_2 - G_2H_1H_2}$$

1.2 Iteration Bound: Data-flow Graph Representations

Many DSP algorithms such as recursive and adaptive digital filters contain feedback loops, which impose an inherent fundamental lower bound on the achievable iteration or sample period. This bound is referred to as the iteration bound or iteration period bound. The iteration bound is a characteristic of the representation of an algorithm in the form of Dataflow Graph (DFG), different representations of the same algorithm may lead to different iteration bounds. It is not possible to achieve an iteration period less than the iteration bound even when infinite processors are available.

DFG Representation

DSP equations are considered to be non-terminating, that run from time index n = 0 to time n = ∞.

For $n = 0 \text{ to } \infty$.

$$y(n) = ay(n-1) + x(n)$$

The input to this is the sequence x(n) for n = 0, 1, 2, ... and the initial condition y(-1). The output is the sequence y(n) for n = 0, 1, 2, ...

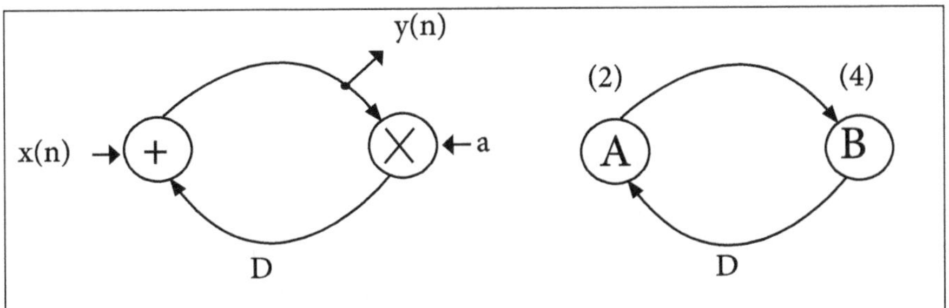

(a) (b)

A graphical representation of y(n) = ay(n - 1) + x(n)A DFG for the given equation.

In the above figure, the numbers in parentheses represent the execution time for the nodes.

A DSP equation is often represented by using a DFG, which is a directed graph that describes the equation. For example, the equation y(n) = ay(n - 1) + x(n) is graphically represented in the above figure (a). A simplified version of this equation is shown in figure (b).

The structure in figure (b) is a DFG, which consists of a set of nodes and edges. The nodes represent the task or computation (Node A represents addition and node B represents multiplication) and each node has an execution time associated with it. The edges represent the communication between the nodes and each edge has a non-negative number of delays associated with it.

In this example, the edge A → B has zero delays and the edge B → A has one delay. An iteration of a node is the execution of the node exactly once and an iteration of the DFG is the execution of each node in the DFG exactly once.

Each edge in a DFG describes a precedence constraint between the two nodes. If the edge has zero delays, then this precedence constraint is an intra-iteration precedence constraint or an inter-iteration precedence constraint, if the edge has one or more delays. Together, the intra-iteration and inter-iteration precedence constraints specify the order in which the nodes in the DFG can be executed.

The edge from A to B in figure (b) enforces the intra-iteration precedence constraint, which states that k^{th} iteration of A must be executed before the kth iteration of B. The edge from B to A enforces the inter-iteration precedence constraint, which states that k^{th} iteration of B must be executed before the $(k + 1)^{th}$ iteration of A.

Let X_k denote the kth iteration of the node X. An intra-iteration precedence constraint is denoted using a single arrow such as $A_k \rightarrow B_k$ and an inter-iteration precedence constraint is denoted using a double arrow such as $B_k \Rightarrow A_k + 1$. Otherwise, all precedence constraints are denoted using a single arrow.

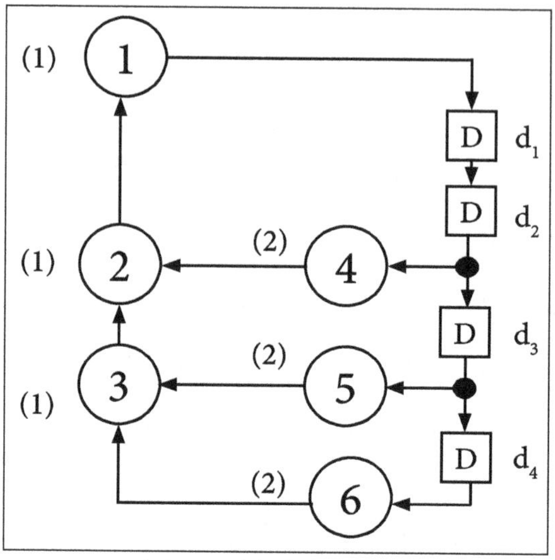

A DFG with three loops.

The iteration bound for this DFG is $T_\infty = 2$ u.t (Normalized units).

The critical path of a DFG is defined to be the path with the longest computation time among all other paths that contain zero delays. The critical path in the DFG of figure (b) is the path A → B, which requires 6 u.t. The DFG in the above figure contains several paths with no delays. The maximum computation time among these paths is 5 u.t. (the paths 6 → 3 → 2 → 1 and 5 → 3 → 2 → 4 → 1 are the critical paths), so the critical path computation requires 5 u.t.

In DFG, the critical path is the longest path for combinational rippling, so the computation time of the critical path is the minimum computation time for one iteration of the DFG.

A DFG can be classified as recursive and non-recursive. A recursive DFG contains at least one loop, whereas a non-recursive DFG contains no loop. For example, an FIR filter is non-recursive, while the DFG in figure (b) is recursive, because it contains the loop A → B → A. A recursive DFG has a fundamental limit on how fast the underlying DSP equation can be implemented in the hardware. This limit, called the iteration bound, T_∞ [1],[2], holds regardless of the computing power available for the implementation of the equation.

1.2.1 Loop Bound and Iteration Bound

A loop is a directed path that begins and ends at the same node. The terms "loop" and "cycle" are used interchangeably to describe a directed cycle in a directed graph. The amount of time required to execute a loop can be determined from the precedence relations described by the edges of the DFG.

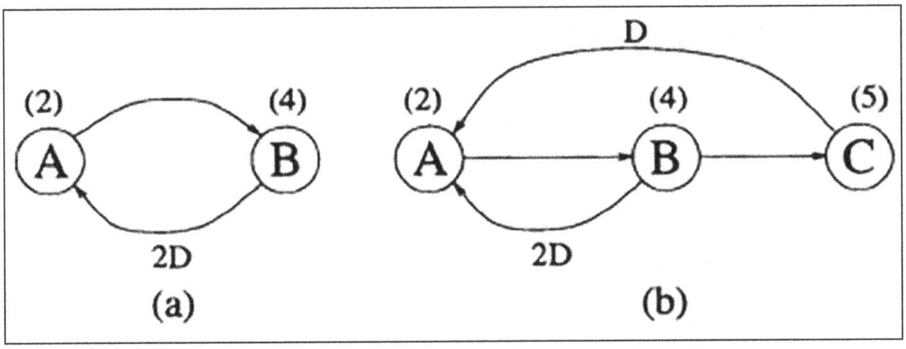

A DFG with one loop A DFG with iteration bound T_∞.

The iteration bound for figure (a) is 3 u.t.

According to these precedence constraints, iteration k of the loop consists of the sequential execution of A_k and B_k. Given that the execution times of nodes A and B are 2 and 4 u.t., respectively, one iteration of the loop requires 6 u.t. This is the loop bound, which represents the lower bound on the loop computation time.

Generally, the loop bound of the lth loop is defined by, t_l / w_l.

Where,

t_l = Loop computation time.

w_l = Number of delays in the loop.

The loop in above figure (a) has two delays, hence it requires 6 u.t. to compute and has a loop bound of 6/2 = 3 u.t. In order to see how one iteration of this loop can be executed in 3 u.t., we need to examine the precedence constraints.

$$A_0 \rightarrow B_0 \Rightarrow A_2 \rightarrow B_2 \Rightarrow A_4 \rightarrow B_4 \Rightarrow A_6 \rightarrow \cdots$$

$$A_1 \rightarrow B_1 \Rightarrow A_3 \rightarrow B_3 \Rightarrow A_5 \rightarrow B_5 \Rightarrow A_7 \rightarrow \cdots$$

This shows the two independent sets of the precedence constraints, one set for the even iterations and one set for the odd iterations. In this case, two iterations can be computed in 6 u.t., resulting in a loop bound of 6/2 = 3.

The DFG in figure (b) contains two loops, namely, the loop $l_1 = A \rightarrow B \rightarrow A$ and $l_2 = A \rightarrow B \rightarrow C \rightarrow A$. The loop bounds for l_1 and l_2 are 6/2 = 3 u.t. and 11/1 = 11 u.t., respectively. The critical loop is the loop with the maximum loop bound, so here, the loop l_2 is the critical loop. The loop bound of the critical loop is the iteration bound of the

DSP equation, which is the lower bound on the iteration or sample period of the DSP equation regardless of the amount of computing resources available.

Generally, the iteration bound is defined as,

$$T_{\infty} = \max_{l \in L} \left\{ \frac{t_l}{w_l} \right\} \qquad \text{...(1)}$$

Where,

L = Set of loops in the DFG.

t_l = Computation time of the loop l.

w_l = Number of delays in the loop l.

For the DFG in figure (b) we have,

$$T_{\infty} = \max \left\{ \frac{6}{2}, \frac{11}{1} \right\} = 11 \text{ u.t} \qquad \text{...(2)}$$

A straightforward technique for finding the iteration bound of a DFG is to locate all the loops and directly compute T_{∞} using equation (1), however, the number of loops in a DFG can be exponential with respect to the number of nodes, so this technique can require long execution times. Three techniques have been developed for computing T_{∞} in polynomial time, namely the longest path matrix algorithm, the minimum cycle mean algorithm and the negative cycle detection algorithm.

1.3 Algorithms for Computing Iteration Bound

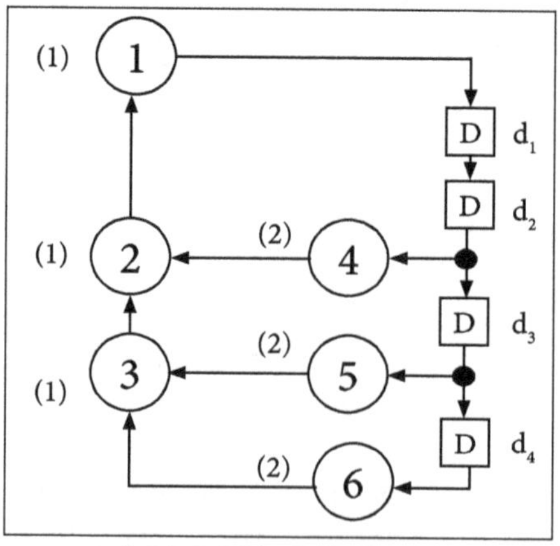

A DFG with three loops

The above DFG has three loops, namely, loop $l_1 = 1 \rightarrow 4 \rightarrow 2 \rightarrow 1$ with loop bound 4/2 u.t., loop $l_2 = 1 \rightarrow 5 \rightarrow 3 \rightarrow 2 \rightarrow 1$ with loop bound 5/3 u.t. and loop $l_3 = 1 \rightarrow 6 \rightarrow 3 \rightarrow 2 \rightarrow 1$ with loop bound 5/4 u.t. Therefore, the iteration bound of this DFG is given by,

$$T_{\infty} = \max\left\{\frac{4}{2}, \frac{5}{3}, \frac{5}{4}\right\} = 2\,\text{u.t.}$$

The two techniques for finding the iteration bound are,

- The Longest Path Matrix Algorithm (LPM).
- The Minimum Cycle Mean Algorithm (MCM).

The Longest Path Matrix Algorithm (LPM)

In the Longest Path Matrix algorithm, a series of matrices are constructed and the iteration bound is found by examining the diagonal elements of the matrices. Let d be the number of delays in the DFG. These matrices, $L^{(m)}$, $m = 1, 2, \ldots, d$, are constructed such that the value of the element $l_{i,j}^{(m)}$ is the longest computation time of all the paths from delay element d_i to the delay element d_j that pass through exactly m - 1 delays. If no such path exists, then $l_{i,j}^{(m)} = -1$.

The longest path between any two nodes can be computed by using any path algorithm. For example, to determine $l_{3,1}^{(1)}$ for the DFG in the above figure, all paths from the delay element d_3 to the delay element d_1 that pass through exactly zero delay elements must be considered. There is one such path, namely, the path $d_3 \rightarrow n_5 \rightarrow n_3 \rightarrow n_2 \rightarrow n_1 \rightarrow d_1$. This path has computation time 5, so $l_{3,1}^{(1)} = 5$. To determine $l_{4,3}^{(1)}$, we note that there are no paths from the delay element d_4 to the delay element d_3 that pass through zero delay elements, so $l_{4,3}^{(1)} = -1$. After determining the rest of the elements of $L^{(1)}$, we find,

$$L^{(1)} = \begin{bmatrix} -1 & 0 & -1 & -1 \\ 4 & -1 & 0 & -1 \\ 5 & -1 & -1 & 0 \\ 5 & -1 & -1 & -1 \end{bmatrix}$$

The higher order matrices, $L^{(m)}$, where, m = 2, 3, ..., d, need not be determined directly from the DFG. Rather, they can be computed recursively according to the rule,

$$l_{i,j}^{(m+1)} = \max_{k \in K}\left(-1, l_{i,k}^{(1)} + l_{k,j}^{(m)}\right)$$

Where, K is a set of integers k in the interval [1, d] such that neither $l_{i,k}^{(1)} = -1$ nor

$l_{k,m}^{(m)} = -1$ holds. For example, to compute $l_{2,1}^{(2)}$, the first step is to find the set K from the possible set {1, 2, 3, 4}. The value 3 is in K, because $l_{2,3}^{(1)} = 0$ and $l_{3,1}^{(1)} = 5$ and the values k = 1, 2, 4 are not in K, because at least one of $l_{2,k}^{(1)}$ or $l_{k,1}^{(1)}$ is equal to -1 for each of these. Using K = {3}, the value of $l_{2,1}^{(2)}$ can be computed as,

$$l_{2.1}^{(2)} = \max_{k \in \{3\}} \left(-1, l_{2,k}^{(1)} + l_{k,1}^{(1)}\right)$$

$$= \max\left(-1, 0 + 5\right) = 5$$

Computing the remaining values of $l_{i,j}^{(2)}$ results in,

$$L^{(2)} = \begin{bmatrix} 4 & -1 & 0 & -1 \\ 5 & 4 & -1 & 0 \\ 5 & 5 & -1 & -1 \\ -1 & 5 & -1 & -1 \end{bmatrix}$$

While $L^{(2)}$ is computed by using only $L^{(1)}$, the matrix $L^{(3)}$ is computed using both $L^{(1)}$ and $L^{(2)}$. To compute $l_{3,3}^{(3)}$ $K = \{1\}$, because $l_{3,1}^{(1)} = 5$ and $l_{1,3}^{(2)} = 0$ and for the values of k = 2, 3, 4, at least one of $l_{3,k}^{(1)}$ or $l_{k,3}^{(3)}$ is equal to −1.

The value of $l_{3,3}^{(3)}$ is given by,

$$l_{3,3}^{(3)} = \max_{k \in \{1\}} \left(-1, l_{3,k}^{(1)} + l_{k,3}^{(2)}\right)$$

$$= \max\left(-1, 5 + 0\right) = 5$$

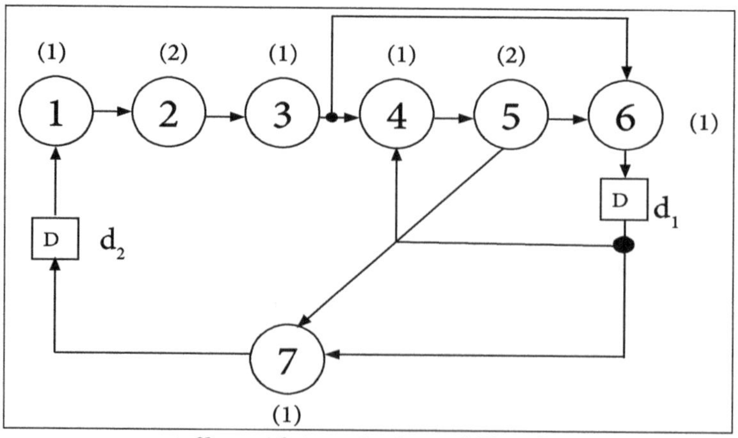

A filter with iteration bound $T_\infty = 8$.

Computing the rest of $L^{(3)}$ and $L^{(4)}$ results in,

$$L^{(3)} = \begin{bmatrix} 5 & 4 & -1 & 0 \\ 8 & 5 & 4 & -1 \\ 9 & 5 & 5 & -1 \\ 9 & -1 & 5 & -1 \end{bmatrix}$$

And,

$$L^{(4)} = \begin{bmatrix} 8 & 5 & 4 & -1 \\ 9 & 8 & 5 & 4 \\ 10 & 9 & 5 & 5 \\ 10 & 9 & -1 & 5 \end{bmatrix}$$

Once the matrices $L^{(m)}$ have been computed, the iteration bound can be determined as,

$$T_\infty = \max_{i,m \in \{1,2,\ldots d\}} \left\{ \frac{l_{i,i}^{(m)}}{m} \right\}$$

Which is,

$$T_\infty = \max\left\{ \frac{4}{2}, \frac{4}{2}, \frac{5}{3}, \frac{5}{3}, \frac{5}{3}, \frac{8}{4}, \frac{8}{4}, \frac{5}{4}, \frac{5}{4} \right\} = 2$$

Consider the DFG of the above figure. The matrices $L^{(m)}$, where, m = 1, 2, are given by,

$$L^{(1)} = \begin{bmatrix} 4 & 4 \\ 8 & 8 \end{bmatrix}$$

And,

$$L^{(2)} = \begin{bmatrix} 12 & 12 \\ 16 & 16 \end{bmatrix}$$

The iteration bound is given by,

$$T_\infty = \max\left\{ \frac{4}{1}, \frac{8}{1}, \frac{12}{2}, \frac{16}{2} \right\} = 8$$

The LPM algorithm works, because the value $l_{i,i}^{(m)}$ represents the longest computation time of all the loops that have m delays and contain the delay element d_i. By taking the maximum of $l_{i,i}^{(m)}/m$ for all possible values of i and m, we find the maximum loop bound of all loops in the DFG, which is the iteration bound.

Since there are elements in $L^{(k+1)}$ and each computation has a time complexity of O(d), the time complexity of computing $L^{(k+1)}$ from $L^{(1)}$ and $L^{(k)}$ is $O\left(d^3\right)$. Therefore, computing $L^{(d)}$ from $L^{(1)}$ has a time complexity of $O\left(d^4\right)$.

Hence, the time complexity of the LPM algorithm for computing the iteration bound is $O\left(d^4 + de\right)$. Note that $L^{(d)}$ can be computed in $O(d^3 \log d)$ complexity by computing $L^{(2)}$ from $L^{(1)}$, $L^{(4)}$ from $L^{(2)}$, $L^{(8)}$ from $L^{(4)}$, etc.

The Minimum Cycle Mean Algorithm (MCM)

The Minimum Cycle Mean algorithm reduces the problem of determining the iteration bound to the problem of finding the MCM of a graph. The technique is then used to efficiently compute the MCM. The terms "loop" and "cycle" can be used interchangeably.

This algorithm uses the concept of a cycle mean, the maximum cycle mean and the MCM. The cycle mean m(c) of a cycle c is the average length of the edges in c, which can be found by taking the sum of the edge lengths and dividing it by the number of edges in the cycle. The MCM λ_{min} is simply the minimum value of all of the cycle means, (i.e.),

$$\lambda_{min} = \min_{c} m(c)$$

Similarly, the maximum cycle mean λ_{max} is given by,

$$\lambda_{max} = \max_{c} m(c)$$

The cycle means of a new graph G_d are used to compute the iteration bound, where G_d can be found from the DFG. If d is the number of delay elements in G, then the graph G_d has d nodes, where each node corresponds to one of the delays in G. The weight w(i, j) of the edge from the node i to node j in G_d is the longest path length among all the paths in G from the delay d_i to the delay d_j that do not pass through any delays. If no zero delay path exists from the delay d_i to the delay d_j, then the edge $i \rightarrow j$ does not exist in G_d. Note that constructing G_d is the same as constructing the matrix $L^{(1)}$ in the LPM algorithm.

The sum of the edge weights in a cycle c in G_d is the maximum computation time of all the cycles in G that contain the delays represented by the nodes in the cycle c. This is because the edge weights in G_d are the maximum computation times between the delays in G. For example, there are two cycles that contain the delays D_α and D_β in the graph G in the below figure (2a) and these cycles have the computation time of 6 u.t. and 4 u.t.

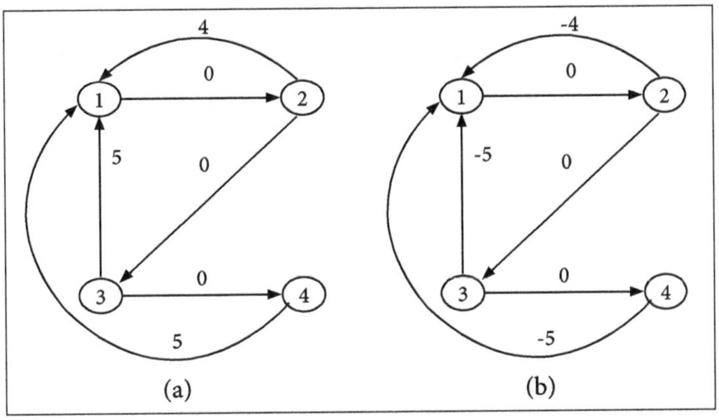

(1) The graphs G_d and $\bar{G}_d$ for the DFG.

The corresponding graph G_d in the below figure (2b) has one cycle that passes through the nodes corresponding to D_α and D_β. The sum of the edge weights in this cycle is 6, which is the maximum computation time of the two cycles in G in the below figure (2a). In G_d, the number of edges in a cycle equals the number of nodes in the cycle and this equals the number of delays in the cycle in G.

Therefore, the cycle mean of a cycle c in G_d is given by,

$$\frac{\text{max computation time of all cycles in G that contain the delays in c}}{\text{the number of delays in these cycles in G}}$$

This is the maximum cycle bound of the cycles in G that contain the delays in the cycle c. The maximum cycle mean of G_d is the maximum cycle bound of all the cycles in G, which is the iteration bound of G.

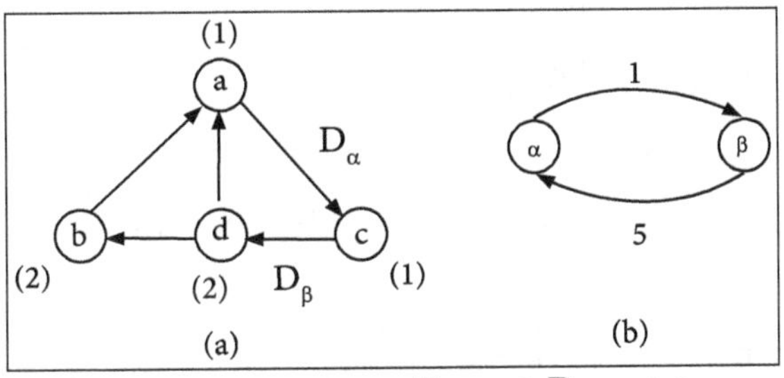

A graph G with two cycles that contain the delays D_α and D_β and the corresponding graph G_d.

To compute the maximum cycle mean of G_d, the graph $\overline{G}_d$ is constructed from G_d by multiplying the weights of the edges by -1, (i.e.) $\overline{G}_d$ has the same topology as G_d and the weights $\bar{w}(i, j)$ of the edge $i \rightarrow j$ in $\overline{G}_d$ are given by, $\bar{w}(i, j) = -w(i, j)$, where $w(i, j)$ is the weight of the edge $i \rightarrow j$ in G_d. The maximum cycle mean of G_d is simply the MCM of $\overline{G}_d$ multiplied by -1. So, the iteration bound of G can be found by finding the MCM of $\overline{G}_d$ and multiplying it by -1.

The MCM of $\overline{G}_d$ is found by constructing the series of d + 1 vectors, $f^{(m)}$, where, m = 0, 1, ..., d, which are of dimension d × 1. An arbitrary reference node is chosen in $\overline{G}_d$. The initial vector $f^{(0)}$ is formed by setting $f^{(0)}(s) = 0$ and then setting the remaining entries of $f^{(0)}$ to ∞. If node 1 is chosen as the reference node for the graph $\overline{G}_d$, then,

$$f^{(0)} = \begin{bmatrix} 0 \\ \infty \\ \infty \\ \infty \end{bmatrix}$$

The remaining vectors, $f^{(m)}$, where, m = 1, 2, ..., d, are recursively computed according to,

$$f^{(m)}(j) = \min_{i \in I}\left(f^{(m-1)}(i) + \overline{w}(i,j)\right)$$

Where,

$\overline{w}$ (i, j) = Weight of the edge i → j in $\overline{G}_d$.

I = Set of nodes in $\overline{G}_d$, such that there exists an edge from node i to node j.

This series of vectors found from $\overline{G}_d$ is,

$$f^{(1)} = \begin{bmatrix} \infty \\ 0 \\ \infty \\ \infty \end{bmatrix} \quad f^{(2)} = \begin{bmatrix} -4 \\ \infty \\ 0 \\ \infty \end{bmatrix} \quad f^{(3)} = \begin{bmatrix} -5 \\ -4 \\ \infty \\ 0 \end{bmatrix} \quad f^{(4)} = \begin{bmatrix} -8 \\ -5 \\ -4 \\ \infty \end{bmatrix}$$

(a) Values of $\left\{\dfrac{f^{(4)}(i) - f^{(m)}(i)}{4-m}\right\}$ for $1 \le i \le 4$ and $0 \le m \le 3$:

	m = 0	m = 1	m = 2	m = 3	$\max_{0 \le m \le 3}\left\{\dfrac{f^{(4)}(i) - f^{(m)}(i)}{4-m}\right\}$
i = 1	−2	−∞	−2	−3	−2
i = 2	−∞	−5/3	−∞	−1	−1
i = 3	−∞	−∞	−2	−∞	−2

$i = 4$	$\infty - \infty$	$\infty - \infty$	$\infty - \infty$	∞	∞

For example, $f^{(4)}(1)$ is computed as,

$$f^{(4)}(1) = \min\left(f^{(3)}(2) - w(2, 1),\ f^{(3)}(3) - w(3, 1),\ f^{(3)}(4) - w(4, 1)\right)$$

$$= \min(-4-4,\ \infty - 5,\ 0-5) = -8$$

From the vectors $f^{(m)}$, where, m = 0, 1, ..., d, the iteration bound can be computed as,

$$T_\infty = -\min_{i\in\{1,2,\ldots,d\}}\left(\max_{m\in\{0,1,\ldots,d-1\}}\left(\frac{f^{(d)}(i) - f^{(m)}(i)}{d-m}\right)\right)$$

Here, d = 4.

In some cases, we may encounter $f^{(d)}(i) - f^{(m)}(i) = \infty - \infty$. In this case, $\infty - \infty$ should be treated as zero. For example, $f^{(4)}(4) - f^{(1)}(4) = \infty - \infty$, so the value $(\infty - \infty)/3 = 0/3 = 0$ is given for i = 4 and m = 1 in table (a).

Using the right column of table (a), we can determine, $T_\infty = -\min\{-2, -1, -2, \infty\} = 2$.

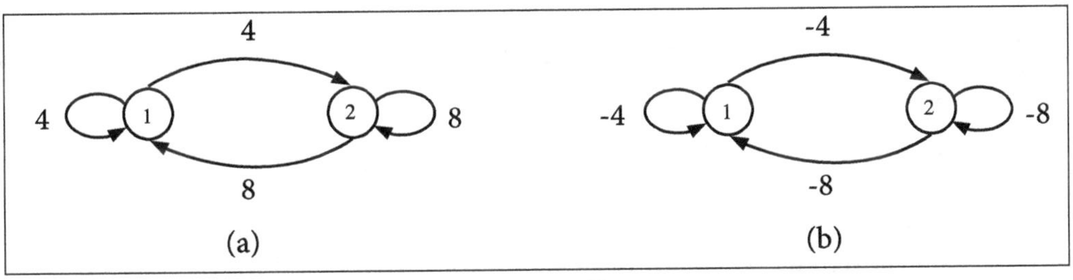

(3) The graphs G_d and $\overline{G}_d$ for the DFG.

(b) Values of $\left\{\frac{f^{(2)}(i) - f^{(m)}(i)}{2-m}\right\}$ for 1 < i < 2 and 0 < m < 1:

	m = 0	m = 1	$\max_{0 \le m \le 1}\left\{\frac{f^{(2)}(i) - f^{(m)}(i)}{2-m}\right\}$
i = 1	– 12/2	– 8/1	– 6
i = 2	$-\infty$	– 8/1	– 8

The set of vectors $f^{(m)}$, where, m = 0, 1, 2, is given by,

$$f^{(0)} = \begin{bmatrix} 0 \\ \infty \end{bmatrix} \quad f^{(1)} = \begin{bmatrix} -4 \\ -4 \end{bmatrix} \quad f^{(2)} = \begin{bmatrix} -12 \\ -12 \end{bmatrix}$$

Using the right column of table (b), we can determine, $T_{\infty} = -\min\{-8,-6\} = 8$.

The time complexity of constructing G_d and $\overline{G}_d$ from the original DFG is O(de). Note that, this is the same as the complexity of computing $L^{(1)}$ in the LPM algorithm, because these are equivalent problems. The MCM of G_d is computed in $O(de_d)$ time, where e_d is the number of edges in G_d. The total time complexity for computing the iteration bound using the MCM algorithm is $O(de + de_d)$.

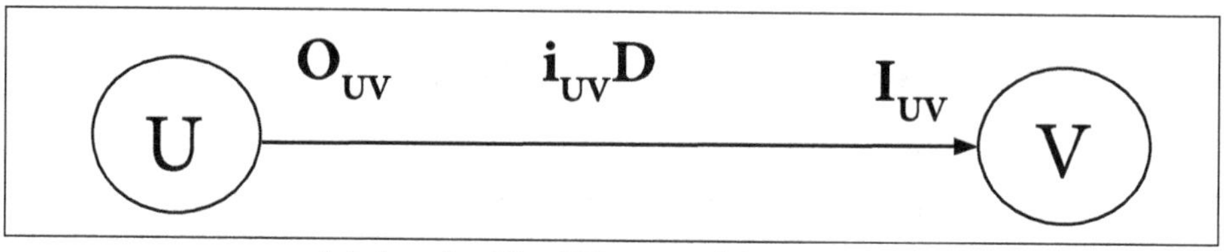

(4) An edge U → V in an multi rate DFG.

1.3.1 Iteration Bound of Multi Rate Data-flow Graphs

Another class of DFGs, called Multi rate DFGs (MRDFGs), allow each node to be executed more than once per iteration and 2 nodes are not required to execute the same number of times in iteration. A 2-step process can be used to compute the iteration bound of a multi rate DFG. This 2-step process is,

- Construct a SRDFG that is equivalent to the MRDFG.
- Compute the iteration bound of the equivalent SRDFG using the LPM or the MCM algorithm.

The iteration bound of the MRDFG is the same as the iteration bound of the equivalent SRDFG.

The value O_{UV} in figure (4) is the number of samples produced on the edge by an invocation of the node U and similarly the value I_{UV} is the number of samples consumed from the edge by an invocation of the node V. The value i_{UV} is the number of delays on the edge.

If the nodes U and V are invoked k_U times and k_V times, respectively, in an iteration, then the number of samples produced on the edge from the node U to node V in one iteration is $O_{UV}k_U$ and then the number of samples consumed from the edge by the node V in one iteration is $I_{UV}k_V$. To avoid a build-up or deficiency of the samples on the edge, the number of samples produced in one iteration must equal the number of samples consumed in one iteration.

This relationship can be described mathematically as,

$$O_{UV}\, k_U = I_{UV}\, k_V \qquad \text{...(1)}$$

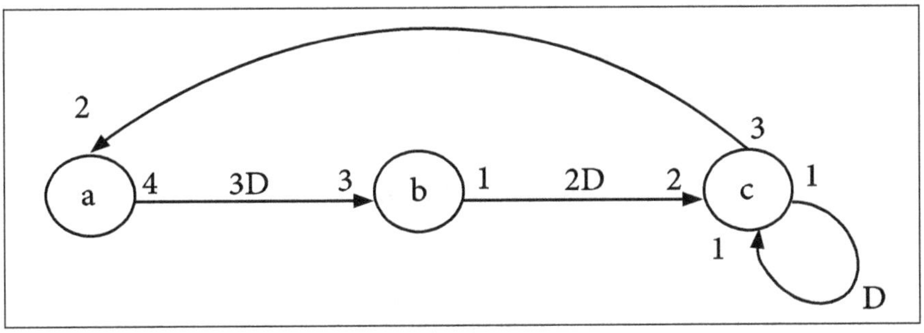

An multi rate DFG.

An algorithm for constructing an equivalent SRDFG from an MRDFG:

For each node U in the MRDFG:

For $k = 0$ to $k_U - 1$

Draw a node U^k in the SRDFG with the same computation time as U in the MRDFG.

For each edge $U \xrightarrow{iUY} V$ in the MRDFG

For $j = 0$ to $O_{UV} k_u - 1$

Draw an edge $U^{j/OUV} \rightarrow V^{(j+i_{UV}/I_{UV})\%K_V}$ in the SRDFG with $(j + i_{UV})/(I_{UV} k_V)$ delays

To determine how many times each node must be executed in iteration, a set of equations found by writing equation (1) for each edge in the MRDFG must be solved, so the numbers of invocations of the nodes are co-prime.

For example, the set of equations for the MRDFG is given by,

$$4k_a = 3k_b$$

$$k_b = 2k_c$$

$$k_c = k_c$$

$$k_c = k_c$$

These equations have a solution, $k_a = 3$, $k_b = 4$ and $k_c = 2$. Once the number of invocations of the nodes are determined, an equivalent SRDFG can be constructed for the MRDFG.

To simplify the notation, let a/b denote the integer part of division and let a%b denote the remainder.

For example, 10/3 = 3 and 10%3 = 1. Mathematically, we can write $a/b = \left\lfloor \frac{a}{b} \right\rfloor$ and a%b

$= a - \left\lfloor \frac{a}{b} \right\rfloor b$, where, $\lfloor x \rfloor$ is the floor of x, which is the largest integer less than or equal to x.

The iteration bound of the SRDFG in the below figure is the same as the iteration bound of the MRDFG:

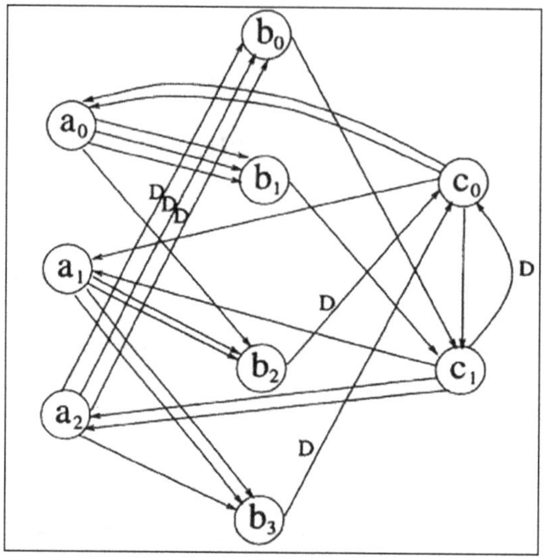

An equivalent SRDFG for the MRDFG.

Various techniques have been suggested to reduce the number of nodes and edges in an equivalent SRDFG without affecting the iteration bound of the graph. These reductions can be used to reduce the complexity of determining the iteration bound of the equivalent SRDFG, which directly leads to reduced complexity of determining the iteration bound of the MRDFG.

1.4 Pipelining and Parallel Processing: Pipelining of FIR Digital Filters

Pipelining transformation leads to a reduction in the critical path, which can be exploited to either increase the clock speed or sample speed or to reduce the power consumption at the same speed. In parallel processing, multiple outputs are computed in parallel in a clock period. Therefore, the effective sampling speed is increased by the level of parallelism. Similar to pipelining, parallel processing can also be used to reduce power consumption.

Consider the 3-tap Finite Impulse Response (FIR) digital filter,

$$y(n) = ax(n) + bx(n-1) + cx(n-2) \qquad ...(1)$$

The block diagram implementation of this filter is shown in the below figure:

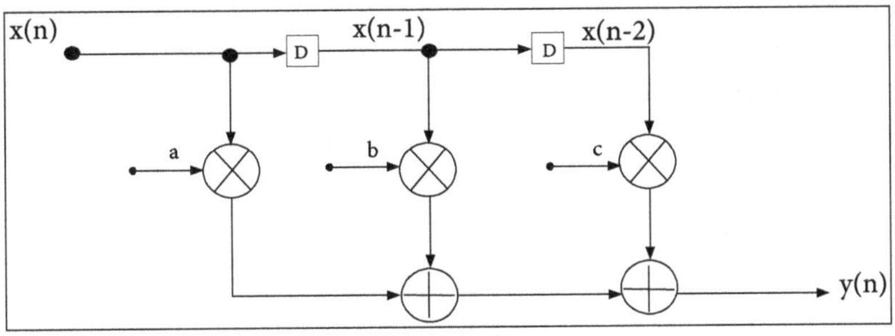

A 3-tap FIR filter.

The critical path or the minimum time required for processing a new sample is limited by 1 multiply and 2 add times, (i.e.) if T_M is the time taken for multiplication and T_A is the time needed for addition operation, then the "sample period" $\left(T_{sample}\right)$ is given by,

$$T_{sample} \geq T_M + 2T_A \qquad ...(2)$$

Therefore, the sampling frequency, $\left(f_{sample}\right)$, also referred to as the throughput or the iteration rate is given by,

$$f_{sample} \leq \frac{1}{T_M + 2T_A} \qquad ...(3)$$

Note that, the direct form structure shown in the above figure can be used only when equation (2) is satisfied, but if some real time application demands a faster input rate, then this structure cannot be used. In that case, the effective critical path can be reduced by using either pipelining processing or parallel processing.

Pipelining reduces the effective critical path by introducing pipelining latches along the data path. Pipelining has been used in the architecture design and compiler synthesis, etc. Parallel processing increases the sampling rate by replicating the hardware, so that several inputs can be processed in parallel and several outputs can be produced at the same time.

Pipelining of FIR Digital Filters

Consider the pipelined implementation of the 3-tap FIR filter of equation (1), obtained by introducing two additional latches as shown in figure (2).

Consider the simple structure in figure (1a), where the computation time of the critical path is $2T_A$. Figure (1b) shows the 2-level pipelined structure, where one latch is placed between the two adders and hence, the critical path is reduced by half. Its 2-level parallel processing structure is shown in figure (1c), where the same hardware is duplicated, so that two inputs can be processed at the same time and two outputs are produced simultaneously. Therefore, the sample rate is increased by 2.

The critical path is now reduced from $T_M + 2T_A$ to $T_M + T_A$. In this arrangement, while the left adder initiates the computation of the current iteration, the right adder completes the computation of the previous iteration result. The schedule of events for this pipelined system is shown in the below table. In this system, at any time, two consecutive outputs are computed in an interleaved manner.

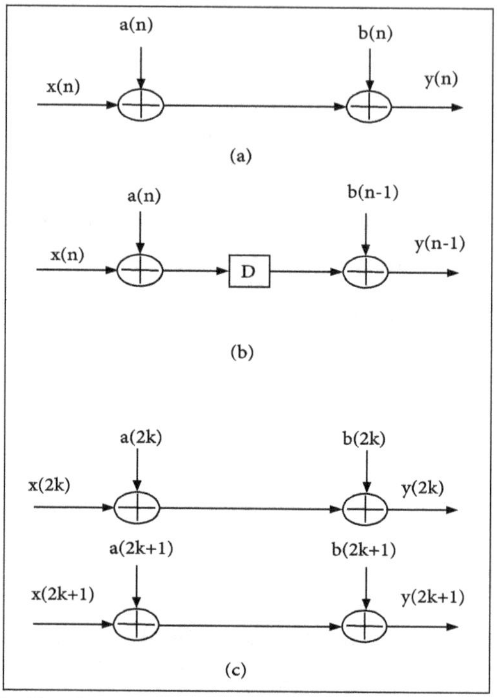

(1) A data path, the 2-level pipelined structure of (a) and the 2-level parallel processing structure of (a).

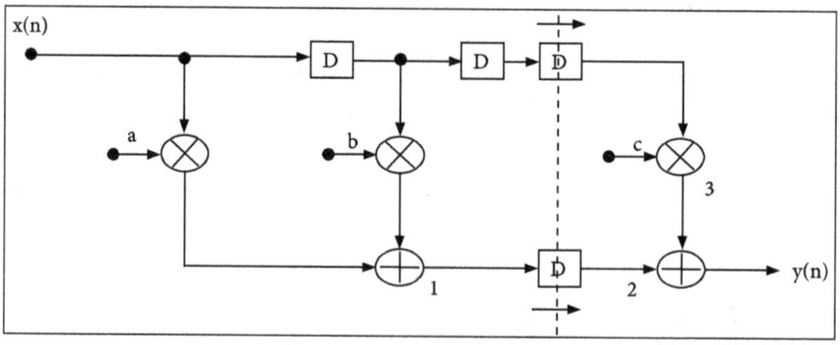

(2) Pipelined FIR filter.

In the above figure, the dashed vertical line represents a feed forward cut set.

Table: Schedule of events in the pipelined FIR filter

Clock	Input	Node 1	Node 2	Node 3	Output
0	x(0)	ax(0) + bx(-1)	-	-	-
1	x(1)	ax(1) + bx(0)	ax(0) + bx(-1)	cx(-2)	y(0)

2	x(2)	ax(2) + bx(1)	ax(1) + bx(0)	cx(-1)	y(1)
3	x(3)	ax(3) + bx(2)	ax(2) + bx(1)	cx(0)	y(2)

Note that in a M-level pipelined system, the number of delay elements in any path from input to output is (M - 1), greater than that in the same path in the original sequential circuit. While pipelining reduces the critical path, it leads to a penalty in terms of an increase in latency.

Latency is the difference in the availability of the first output data in the pipelined system and sequential system. For example, if the latency is 1 clock cycle, then the kth output is available in the $(k + 1)^{th}$ clock cycle in a 1-stage pipelined system.

The two main drawbacks of pipelining are:

- Increase in the number of latches.
- Increase in system latency.

The following points must be noted:

- The speed of architecture or the clock period is limited by the longest path between any two latches or between an input and a latch or between a latch and an output or between the input and the output.
- This longest path or the critical path can be reduced by suitably placing the pipelining latches in the architecture.
- The pipelining latches can be placed only across any feed forward cutset of the graph.

Cut Set: A cut set is a set of edges of a graph, such that if these edges are removed from the graph, the graph becomes disjoint.

Feed Forward Cut Set: A cut set is called a feed forward cut set, if the data move in the forward direction on all the edges of the cut set. For example, the cut set used to pipeline the FIR filter in figure (2) is a feed forward cut set. We can arbitrarily place the latches on a feed forward cut set of any FIR filter structure without affecting the functionality of the algorithm.

1.4.1 Parallel processing

Parallel processing and pipelining techniques are duals of each other and if a computation can be pipelined, it can also be processed in parallel. Both techniques exploit concurrency available in the computation in different ways. While independent sets of computations are computed in an interleaved manner in a pipelined system, they are computed using duplicate hardware in parallel processing mode.

Designing a parallel FIR system:

Consider the 3-tap FIR filter system. This system is a Single-Input Single-Output (SISO) system and is 2described by,

$$y(n) = ax(n) + bx(n-1) + cx(n-2) \qquad ...(1)$$

To obtain a parallel processing structure, the SISO system must be converted into a MIMO (Multiple-Input Multiple-Output) system. For example, the following set of equations describe a parallel system with three inputs per clock cycle (i.e.) level of parallel processing, L = 3.

$$y(3k) = ax(3k) + bx(3k-1) + cx(3k-2) \qquad ...(2)$$

$$y(3k+1) = ax(3k+1) + bx(3k) + cx(3k-1)$$

$$y(3k+2) = ax(3k+2) + bx(3k+1) + cx(3k)$$

Here, k denotes the clock cycle. At the kth clock cycle, the three inputs x(3k), x(3k + 1) and x(3k + 2) are processed and three samples are generated at the output. Parallel processing systems are also referred to as the block processing systems and the number of inputs processed in a clock cycle is referred to as the block size, because of the MIMO structure, placing a latch at any line produces an effective delay of L clock cycles at the sample rate.

Each delay element is referred to as the block delay or L-slow. For example, delaying the signal x(3k) by 1 clock cycle would result in x(3k - 3) instead of x(3k - 1), which has been input in another input line. The block architecture for a 3-parallel FIR filter is shown in the below figure (1) and its details are shown in figure (2).

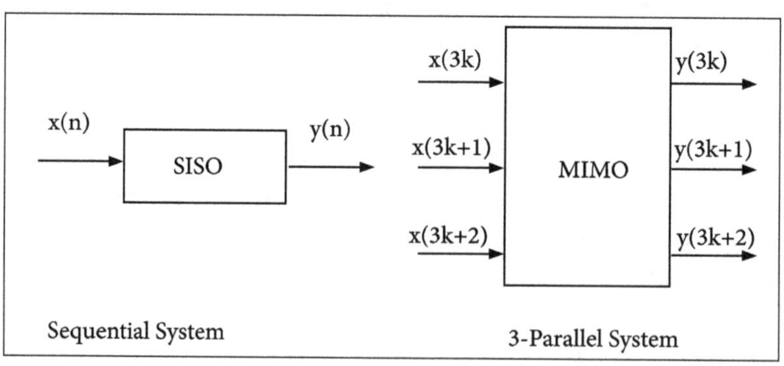

(1) A block processing example.

Note that the critical path of the block or parallel processing system has remained unchanged and the clock period (T_{clk}) must satisfy the following condition:

$$T_{clk} \geq T_M + 2T_A$$

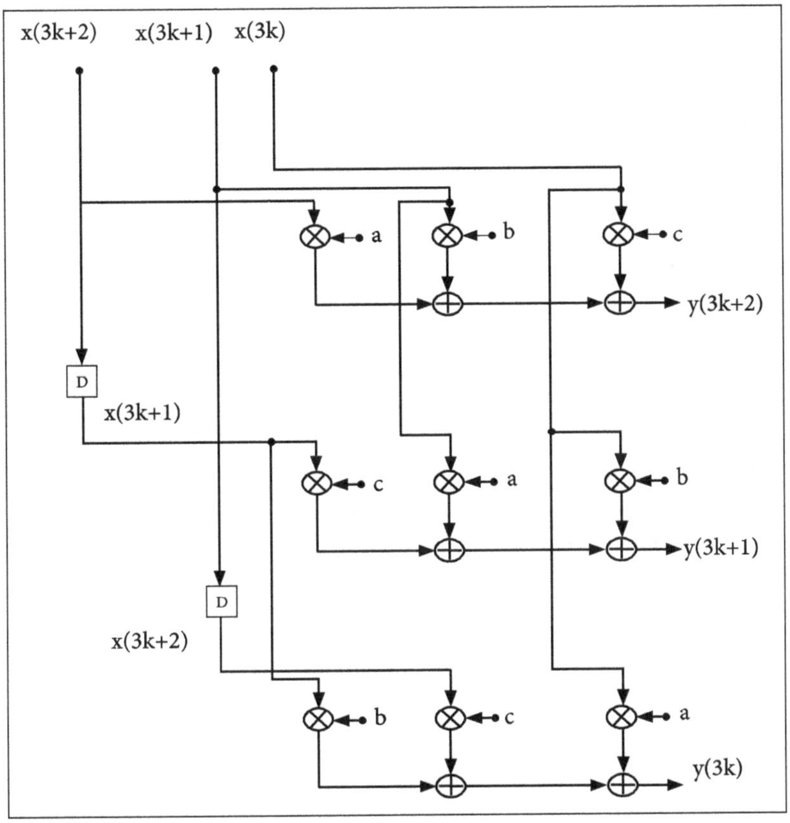

(2) Parallel processing architecture for a 3-tap FIR filter with block size 3.

Since three samples are processed in 1 clock cycle instead of 3, the iteration period is given by,

$$T_{iter} = T_{sample} = \frac{1}{L}T_{clk} \geq \frac{1}{3}(T_M + 2T_A)$$

In a parallel system, $T_{clk} \neq T_{sample}$, whereas in a pipelined system $T_{clk} = T_{sample}$. Figure (3) shows a complete parallel processing system including serial-to-parallel and parallel-to-serial converters, shown in figure (4) and figure (5), respectively.

Consider the chip set shown in figure (6). For example, if output-pad delay plus input-pad delay and the wire delay between the two chips is 8 ns, then T_{clk} has to be greater than or equal to 8 ns. If the critical path computation time is less than 8 ns, then the I/O bound dominates and this system is communication bounded.

This means that, pipelining can be used only to a certain extent such that the critical path computation time is limited by the communication or I/O bound and once this has been reached, pipelining can no longer increase the speed. At this point, pipelining can be combined with parallel processing to further increase the speed of the architecture.

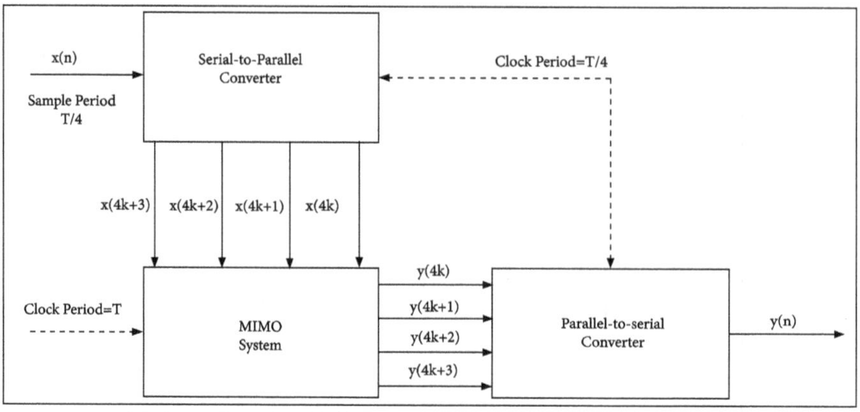

(3) Complete parallel processing system with block size 4.

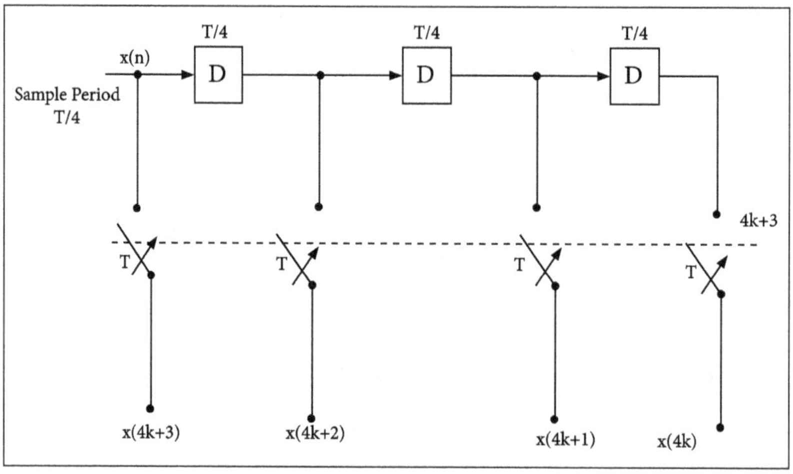

(4) A serial-to-parallel converter.

As an example, consider the parallel filter in figure (2). Assume the computation time for 1 multiplication (T_M) to be 10 u.t. and the computation time for 1 adder (T_A) to be 2 u.t. Fine grain pipelining can be applied to the parallel filter to further reduce the critical path. In this case, the multiplier is broken up into two smaller units, m1 and m2, with computation time 7 u.t. and 3 u.t., respectively and pipelining latches are placed on the horizontal cut sets across the multipliers as shown in figure (7).

These horizontal cut sets may appear to be invalid, however, they are valid, since cutting the edges of these cut sets will lead to disjoint components. Therefore, by combining parallel processing and pipelining, the sample period has been reduced to,

$$T_{iter} = T_{sample} = \frac{1}{LM} T_{clk} = \frac{1}{6}\left(T_M + 2T_A\right)$$

Parallel processing is also used to reduce power consumption while using slow clocks. This reduces the power consumption due to the clock lines, as compared with a pipelined system, which needs to be operated using a faster clock for equivalent throughput or sample speed.

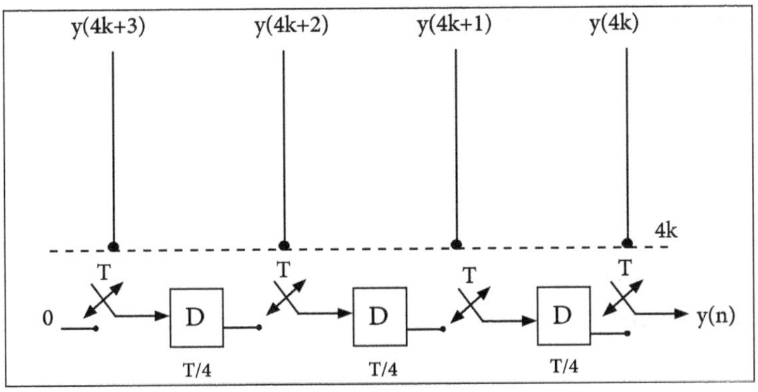

(5) A parallel-to-serial converter.

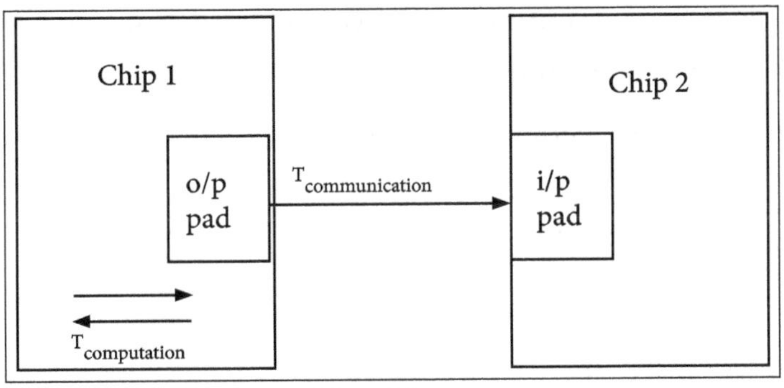

(6) A chip set.

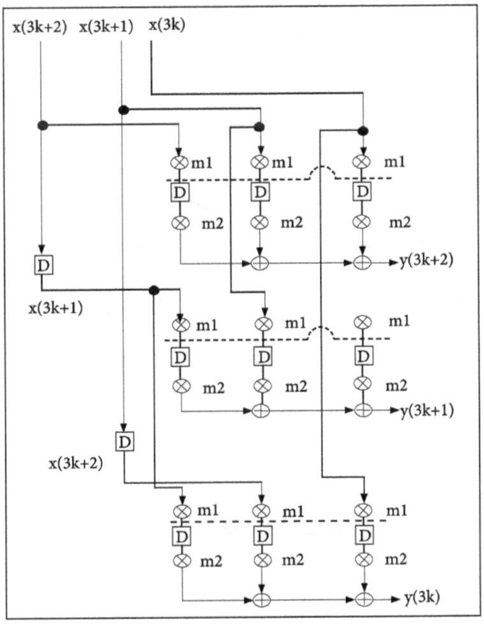

(7) Combined fine-grain pipelining and parallel processing for 3-tap FIR filter.

Furthermore, use of fine grain pipelining such as bit-level pipelining may not be desirable, since the latency increase and overhead hardware due to latches may be significant.

1.4.2 Pipelining and Parallel Processing for Low Power

There are two main advantages of using pipelining and parallel processing. They are,

- Higher speed
- Lower power

Pipelining and parallel processing can increase the sample speed. Now, consider the use of these techniques for lowering the power consumption, where sample speed need not be increased.

Two formulas must be considered, one for computing the propagation delay of CMOS circuits and the other for computing the power consumption. The propagation delay $_{pd}$ is associated with charging and discharging of various gates and stray capacitances in the critical path. For CMOS circuits, the propagation delay can be written as,

$$T_{pd} = \frac{C_{charge} V_o}{k(V_o - V_t)^2} \qquad ...(1)$$

Where,

C_{charge} = Capacitance to be charged/discharged in a single clock cycle, (i.e.) the capacitance along the critical path

V_o = Supply voltage

V_t = Threshold voltage

k = Function of technology parameters such as μ, (W/L) and C_{ox}.

The power consumption of a CMOS circuit can be estimated using the following equation:

$$P = C_{total} V_o^2 f \qquad ...(2)$$

Where,

C_{total} = Total capacitance of the circuit

V_o = Supply voltage

f = Clock frequency of the circuit

Equations (1) and (2) are based on simple approximations and are appropriate only for 1st order analysis.

Pipelining for Low Power

Pipelining can be used to reduce the power consumption of an FIR filter.

$$P_{seq} = C_{total} V_o^2 f \qquad ...(3)$$

Let, equation (3) represent the power consumed in an original filter. Note that, $f = (1/T_{seq})$, where T_{seq} is the clock period of the original sequential filter. Now, consider an M-level pipelined system, where the critical path is reduced to (1/M) of its original length and the capacitance to be charged/discharged in a single clock cycle is reduced to (C_{charge}/M).

Note that, the total capacitance does not change. If the same clock speed is maintained, (i.e.) the clock frequency f is maintained, then only a fraction of the original capacitance, (C_{charge}/M), is being charged/discharged in the same amount of time that was previously needed to charge/discharge the capacitance, C_{charge}. This implies that, the supply voltage can be reduced to βV_o, where β is a positive constant less than 1.

Hence, the power consumption of the pipelined filter will be,

$$P_{pip} = C_{total} \beta^2 V_o^2 f = \beta^2 P_{seq} \qquad ...(4)$$

Therefore, the power consumption of the pipelined system has been reduced by a factor of β^2 as compared with the original system.

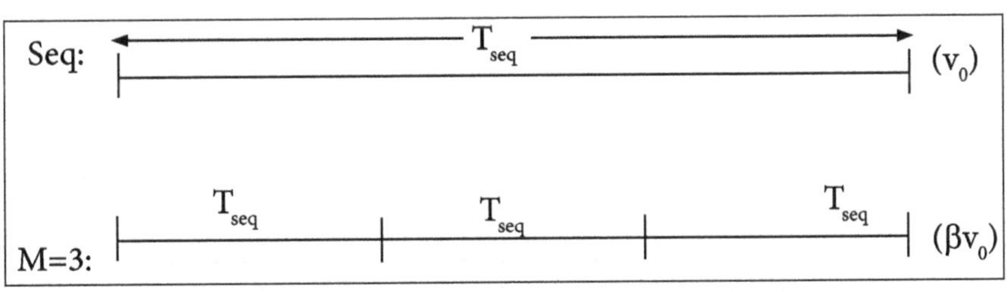

Critical path lengths for original and 3-level pipelined systems.

The power consumption reduction factor, β, can be determined by examining the relationship between the propagation delay of the original filter and the pipelined filter.

The propagation delay of the original filter is given by,

$$T_{seq} = \frac{C_{charge} V_o}{k(V_o - V_t)^2} \qquad ...(5)$$

The propagation delay of the pipelined filter is given by,

$$T_{pip} = \frac{\frac{C_{charge}}{M} \beta V_o}{k(\beta V_o - V_t)^2} \qquad ...(6)$$

Generally, the clock period, T_{clk}, is set equal to the maximum propagation delay, T_{pd}, in a circuit. Since the same clock speed is maintained for both the filters, from equation (5) and (6), the following quadratic equation can be obtained to solve for β:

$$M(\beta V_o - V_t)^2 = \beta(V_o - V_t)^2 \qquad ...(7)$$

Once β is obtained, the reduced power consumption of the pipelined filter can be computed using equation (4).

Parallel Processing for Low Power

Like pipelining, parallel processing can reduce the power consumption of a system by allowing the supply voltage to be reduced. In an L-parallel system, the charging capacitance does not change, while the total capacitance is increased by L times.

In order to maintain the same sample rate, the clock period of the L-parallel circuit must be increased to LT_{seq}, where T_{seq} is the propagation delay of the sequential circuit given by equation (5). This means that C_{charge} is charged in time LT_{seq} rather than in time T_{seq}. In other words, there is more time to charge the same capacitance. This means that the supply voltage can be reduced to βV_o.

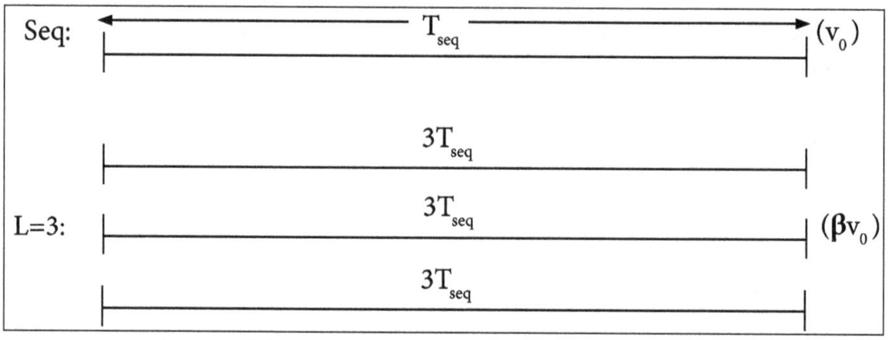

Critical path lengths for sequential and 3-parallel systems.

The propagation delay considerations can be used to compute the supply voltage of the L-parallel system. The propagation delay of the original system is given by (5), while the propagation delay of the L-parallel system is given by,

$$LT_{seq} = \frac{C_{charge}\,\beta V_o}{k(\beta V_o - V_t)^2} \qquad ...(8)$$

From equations (5) and (8), the following quadratic equation can be obtained to compute β:

$$L(\beta V_o - V_t)^2 = \beta(V_o - V_t)^2 \qquad ...(9)$$

Once β is computed, the power consumption of the L-parallel system can be calculated as,

$$P_{par} = (LC_{charge})(\beta V_o)^2 \frac{f}{L} \qquad ...(10)$$

$$= \beta^2 C_{charge} V_o^2 f$$

$$= \beta^2 P_{seq}$$

Where, P_{seq} is the power consumption of the sequential system given by equation (3). Therefore, as in the pipelined system, the power consumption of the L-parallel system has been reduced by a factor of β^2 as compared with the original sequential system.

1.5 Retiming: Definitions and Properties

Retiming is a transformation technique used to change the locations of delay elements in a circuit without affecting the input/output characteristics of the circuit. For example, consider the IIR filter in figure (1a).

This filter is described by,

$$w(n) = ay(n-1) + by(n-2)$$

$$w(n) = ay(n-1) + by(n-2)$$

$$= ay(n-2) + by(n-3) + x(n)$$

The filter in figure (1b) is described by,

$$w_1(n) = ay(n-1)$$

$$w_2(n) = by(n-2)$$

$$y(n) = w_1(n-1) + w_2(n-1) + x(n)$$

$$= ay(n-2) + by(n-3) + x(n)$$

Although the filters in figure (1a) and (1b) have delays at different locations, these filters have the same input/output characteristics. These two filters can be derived from one another using retiming.

Retiming has many applications in synchronous circuit design. These applications include reducing the clock period of the circuit, reducing the number of registers in the circuit, reducing the power consumption of the circuit and logic synthesis.

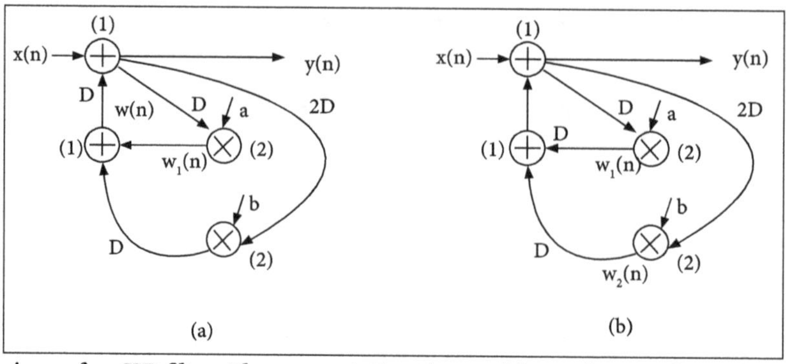

(1) Two versions of an IIR filter. The computation times of the nodes are shown in parentheses.

Retiming can be used to increase the clock rate of a circuit by reducing the computation time of the critical path. The critical path is defined to be the path with the longest computation time among all the paths that contain zero delays and the computation time of the critical path is lower bound on the clock period of the circuit.

The critical path of the filter in figure (1a) passes through one multiplier and one adder and has a computation time of 3 u.t., so this filter cannot be clocked with a clock period of less than 3 u.t. The retimed filter in figure (1b) has a critical path that passes through two adders and has a computation time of 2 u.t., so this filter can be clocked with a clock period of 2 u.t. By retiming the filter in figure (1a) to obtain the filter in figure (1b), the clock period has been reduced from 3 u.t. to 2 u.t. or by 33%.

Retiming can be used to decrease the number of registers in a circuit. The filter in figure (1a) uses four registers, while the filter in figure (1b) uses five registers. Since retiming can affect the clock period and the number of registers, it is sometimes desirable to take both of these parameters into account.

Retiming can be used to reduce the power consumption of a circuit by reducing switching, which can lead to dynamic power dissipation in static CMOS circuits. Placing registers at the inputs of nodes with large capacitances can reduce the switching activities at these nodes, which can lead to low power solutions.

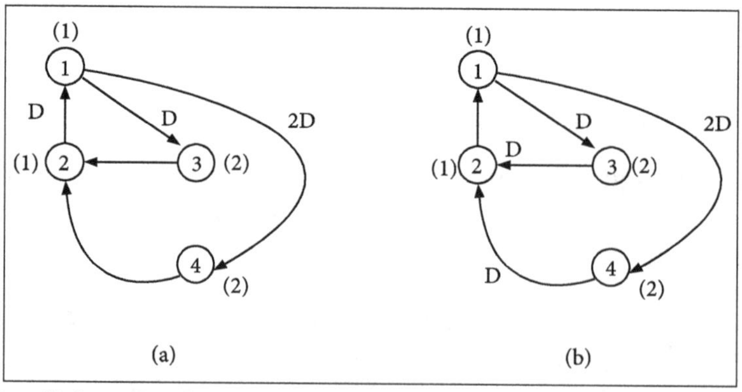

(2) A DFG and The retimed DFG obtained using r(1) = 0, r(2) = 1, r(3) = 0 and r(4) = 0.

Definitions and Properties

Retiming maps a circuit G to a retimed circuit G_r. A retiming solution is characterized by a value r(V) for each node V in the graph. Let w(e) denote the weight of the edge e in the original graph G and let $w_r(e)$ denote the weight of the edge e in the retimed graph G_r.

The weight of the edge $U \xrightarrow{e} V$ in the retimed graph is computed from the weight of the edge in the original graph using,

$$w_r(e) = w(e) + r(V) - r(U) \quad ...(1)$$

To demonstrate some retiming concepts, the filter in figure (1a) is redrawn in figure (2a) and the retimed filter in figure (1b) is redrawn in figure (2b). The retiming values, r(1) = 0, r(2) = 1, r(3) = 0 and r(4) = 0 can be used to obtain the retimed DFG in figure (2b) from the DFG in figure (2a).

For example, the edge $3 \xrightarrow{e} 2$ in the retimed DFG contains,

$$w_r(3 \xrightarrow{e} 2) = w(3 \xrightarrow{e} 2) + r(2) - r(3)$$

= 0 + 1 - 0 = 1 delay

Similarly, the edge $2 \xrightarrow{e} 1$ contains,

$$w_r(2 \xrightarrow{e} 1) = w(2 \xrightarrow{e} 1) + r(1) - r(2)$$

= 1 + 0 - 1 = 0 delays

A retiming solution is feasible, if $w_r(e) \geq 0$ holds for all edges. While the solution that maps figure (2a) to figure (2b) is feasible, because all the edges in figure (2b) have non-negative weights, the solution r(1) = 0, r(2) = -1, r(3) = 0 and r(4) = 0 is infeasible because, for example, the edge $3 \xrightarrow{e} 2$ in the retimed system contains,

$$w_r(3 \xrightarrow{e} 2) = w(3 \xrightarrow{e} 2) + r(2) - r(3)$$

= 0 + (-1) - 0 = -1 delays

Properties of Retiming

Several properties of retiming can be derived from the retiming equation (1). A path is a sequence of edges and nodes $V_0 \xrightarrow{e_0} V_1 \xrightarrow{e_1} \cdots \xrightarrow{e_{k-2}} V_{k-1} \xrightarrow{e_{k-1}} V_k$.. The weight of the path p is $w(p) = \sum_{i=0}^{k-1} w(e_i)$ and the computation time of the path is $t(p) = \sum_{i=0}^{k} t(V_i)$. A cycle is a closed path $V_0 \xrightarrow{e_0} V_1 \xrightarrow{e_1} \cdots \xrightarrow{e_{k-2}} V_{k-1} \xrightarrow{e_{k-1}} V_0$.

Property-1

The weight of the retimed path $p = V_0 \xrightarrow{e_0} V_1 \xrightarrow{e_1} \cdots \xrightarrow{e_{k-1}} V_k$ is given by, $w_r(p) = w(p) + r(V_k) - r(V_0)$

The retimed path weight is given by,

$$
\begin{aligned}
w_r(p) &= \sum_{i=0}^{k-1} w_r(e_i) \\
&= \sum_{i=0}^{k-1} \left(w(e_i) + r(V_{i+1}) - r(V_i)\right) \\
&= \sum_{i=0}^{k-1} w(e_i) + \left(\sum_{i=0}^{k-1} r(V_{i+1}) - \sum_{i=0}^{k-1} r(V_i) \right) \\
&= w(p) + r(V_k) - r(V_0)
\end{aligned}
$$

For example, the path 2 → 1 → 3 in figure (2a) has two delays and this path in the re-timed DFG in figure (2b) and has 2 + r(3) - r(2) = 2 + 0 - 1 = 1 delay.

Property-2

Retiming does not change the number of delays in a cycle.

This is a special case of property (1), where, $V_k = V_0$. The weight of the retimed cycle c is, $w_r(c) = w(c) + r(V_0) - r(V_0) = w(c)$. In figure (2), the cycle 1 → 3 → 2 → 1 contains 2 delays in the unretimed and retimed DFGs and the cycle 1 → 4 → 2 → 1 contains 3 delays in the unretimed and retimed DFGs.

Property-3

Retiming does not alter the iteration bound in a DFG.

Property-4

Adding the constant value j to the retiming value of each node does not change the mapping from G to G_r.

After replacing r(V) with r(V) + j for each node, the weight of the retimed edge $U \xrightarrow{e} V$ in G_r is given by,

$$w_r(e) = w(e) + \left(r(V) + j\right) - \left(r(U) + j\right) = w(e) + r(v) - r(U)$$

This is the same for any value of j (including j = 0). The retiming values r(1) = 0, r(2) = 1, r(3) = 0 and r(4) = 0 are used to obtain the retimed DFG in figure (2b) from the unretimed DFG in figure (2a). For example, by adding the constant -38 to these retiming values, the retiming values r(1) = -38, r(2) = -37, r(3) = -38 and r(4) = -38 can be used to obtain the retimed DFG in figure (2b) from the DFG in figure (2a).

1.5.1 Solving Systems of Inequalities

Given a set of M inequalities in N variables, where each inequality has the form $r_i - r_j \le k$ for integer values of k, one of the shortest path algorithms can be used to determine if a solution exists and to find a solution if one does really exist.

This is done using the following procedure:

1. Draw a constraint graph,

- Draw the node i for each of the N variables r_i, where, i = 1, 2, ..., N.
- Draw the node N + 1.
- For each inequality, $r_i - r_j \le k$, draw the edge j → i from the node j to the node i with length k.
- For each node i, where, i = 1, 2, ..., n, draw the edge N + 1 → i from the node N + 1 to the node i with length 0.

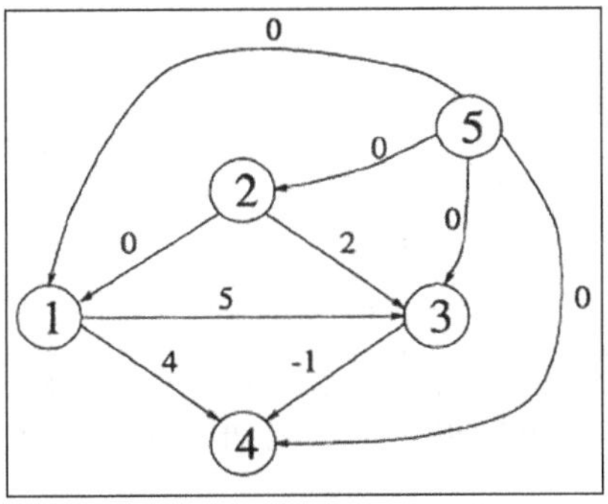

(3) The constraint graph.

2. Solve using a shortest path algorithm,

- The system of inequalities has a solution, if and only if the constraint graph contains no negative cycles.
- If a solution exists, one solution is where r_i is the minimum length path from the node N + 1 to node i.

1.5.2 Retiming Techniques

First, two special cases of retiming, namely, cut set retiming and pipelining, are considered. Two algorithms are then considered for retiming to minimize the clock period and retiming to minimize the number of registers that are required to implement the circuit.

Cut Set Retiming and Pipelining

Cut set retiming is a useful technique that is a special case of retiming. A cut set is a set of edges that can be removed from the graph to create two disconnected sub-graphs. Cut set retiming only affects the weights of the edges in the cut set. If the two disconnected sub-graphs are labeled G_1 and G_2, then cut set retiming consists of adding k delays to each edge from G_1 to G_2 and removing k delays from each edge from G_2 to G_1.

For example, a cut set is shown with a dashed line in figure (4a). The three edges in the cut set are 2 → 1, 3 → 2 and 1 → 4. The two sub-graphs G_1 and G_2 found by removing the three edges in the cut set are shown in figure (4b). For k = 1, the result of cut set retiming is shown in figure (4c). The edges from G_1 to G_2 are 3 → 2 and 1→ 4 and one delay is added to each of these edges. The edge from G_2 to G_1 is 2 → 1 and one delay is subtracted from this edge.

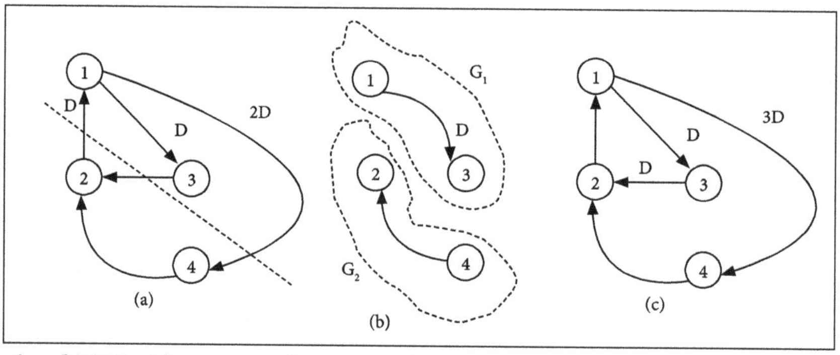

(4) The unretimed DFG with a cut set shown as a dashed line, The two graphs G_1 and G_2 formed by removing the edges in the cut set and The retimed graph found using cut set retiming with k = 1.

Cut set retiming is a special case of retiming where each node in the sub graph G_1 has the retiming value j and each node in the sub graph G_2 has the retiming value j + k. The value of j is not important. For example in figure (4), using the values j = 0 and k = 1 results in r(1) = 0, r(2) = 1, r(3) = 0 and r(4) = 1 and this maps the DFG in figure (4a) to the DFG in figure (4c). Any value of j results in the same retimed graph.

For feasibility of the retimed graph, $w_r(e) \geq 0$ must hold for all edges e in G_r. Let $e_{1,2}$ represent an edge from G_1 to G_2 and let $e_{2,1}$ represent an edge from G_2 to G_1. Since cut set retiming adds k delays to each edge from G_1 to G_2, $w_r(e_{1,2}) \geq 0 \Rightarrow w(e_{1,2}) + k \geq 0$ must hold. Similarly, since k delays are subtracted from each edge $e_{2,1}$ from G_2 to G_1, $w_r(e_{2,1}) \geq 0 \Rightarrow w(e_{2,1}) - k \geq 0$ must hold.

Combining these two inequalities and considering all of the edges in the cut set result in,

$$-\min_{G_1 \xrightarrow{e} G_2} \{w(e)\} \le k \le \min_{G_2 \xrightarrow{e} G_1} \{w(e)\}$$

as the condition on k for cut set retiming to give a feasible solution.

In figure (4), $\min_{G_1 \xrightarrow{e} G_2} \{w(e)\} = \min\{0,2\} = 0$ and $\min_{G_2 \xrightarrow{e} G_1} \{w(e)\} = 1$, so $0 \le k \le 1$ must hold. Since k = 0 does nothing, k = 1 is the only value of k that results in a feasible retimed graph and this graph is shown in figure (4c).

An example of such a cut set is shown in figure (5a) and the two sub-graphs G_1 and G_2 are shown in figure (5b). The graph G_2 consists of the node 2 and the graph G_1 consists of the rest of the graph except the three edges that go into and out of the node 2.

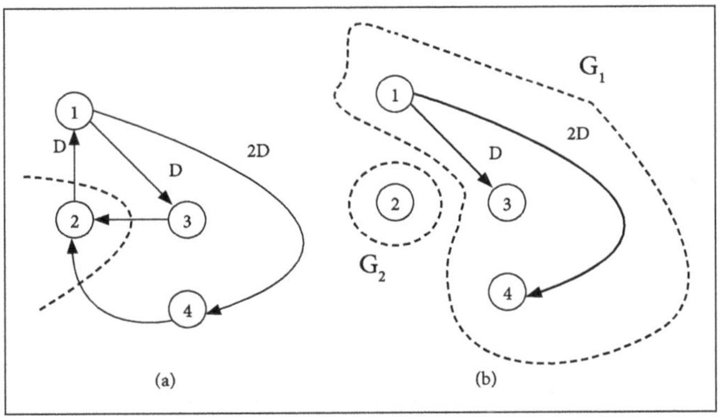

(5) The unretimed DFG with a cut set shown as a dashed line and The two graphs G1 and G2 formed by removing the edges in the cut set.

The result of cut set retiming in this case is found by adding 1 delay to each edge incident into the node 2 and subtracting 1 delay from each edge outgoing from the node 2. Therefore, this special case of cut set retiming consists of choosing a node as a cut set and subtracting one delay from each edge outgoing from the node and adding one delay to each edge incident into the node.

Pipelining is a special case of cut set retiming where there are no edges in the cut set from the sub-graph G_2 to the sub-graph G_1, (i.e.) pipelining applies to graphs without loops. These cut sets are referred to as feed-forward cut sets. The feed-forward cut set shown in figure (6a) divides the graph into the two sub-graphs shown in figure (6b).

All three of the edges in the cut set go from G_1 to G_2 and performing cut set retiming with k = 2 results in two additional delays on each edge in the cut set, resulting in the retimed graph in figure (6c). This demonstrates that pipelining can be viewed as a special case of retiming.

Cut set retiming is often used in combination with slow-down. The procedure is to first replace each delay in the DFG with N delays to create an N-slow version of the DFG and then to perform cut set retiming on the N-slow DFG. Note that in an N-slow system, N - 1 null operation must be interleaved after each useful signal sample to preserve the functionality of the algorithm.

For example, consider the 100-stage lattice filter in figure (7a), which has a critical path of 101 adders and two multipliers. Assuming that addition and multiplication take 1 and 2 u.t., respectively, the minimum sample period is (101)(1) + (2)(2) = 105 u.t. The 2-slow version of this circuit is shown in figure (7b) and cut set retiming can be used to obtain the circuit in figure (7c).

The critical path of the retimed circuit has 2 adders and 2 multipliers and has computation time (2)(1) + (2)(2) = 6 u.t. Since this circuit is 2-slow, the minimum sample period is (2)(6) = 12 u.t. In this example, slow-down and cut set retiming reduce the sample period from 105 u.t. to 12 u.t.

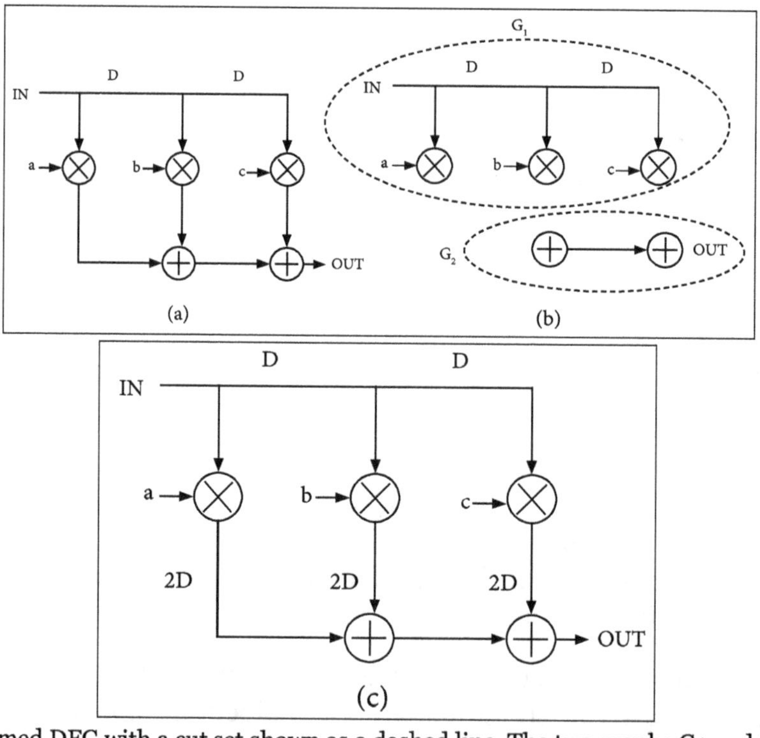

(6) The unretimed DFG with a cut set shown as a dashed line, The two graphs G1 and G2 formed by removing the edges in the cut set and The graph obtained by cut set retiming with k = 2.

Cut set retiming is a special case of retiming and pipelining is a special case of cut set retiming. Cut set retiming and pipelining are graphical techniques that can be used to perform complex retiming operations in a simple manner.

Retiming for Clock Period Minimization

For a circuit G, the minimum feasible clock period is the computation time of the critical path, which is the path with the longest computation time among all paths with no delays. Mathematically, the minimum feasible clock period, φ(G), is defined as,

$$\phi(G) = \max\{t(p) : w(p) = 0\}$$

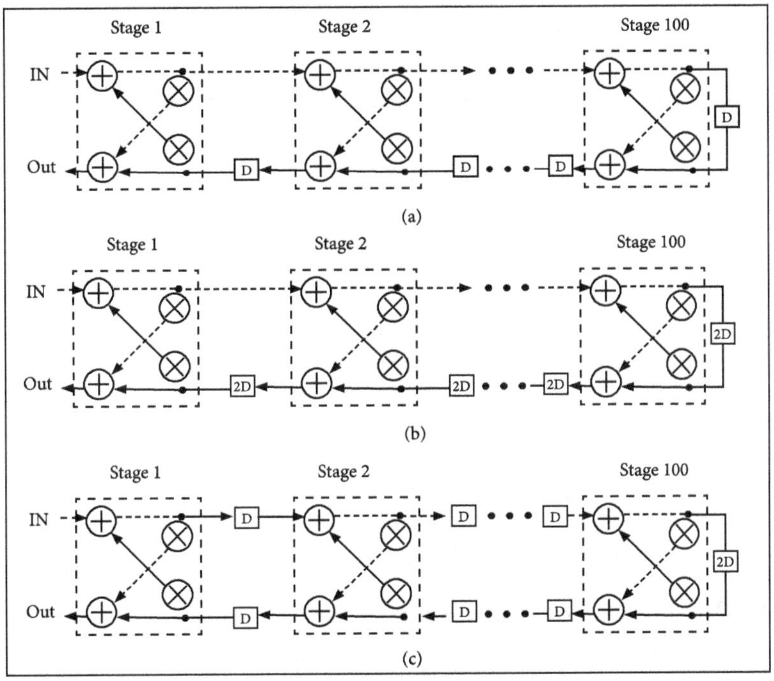

(7) A 100-stage lattice filter with minimum sample period of 105 u.t, The 2-slow version of the circuit and A retimed version of the 2-slow circuit with critical path of 6 u.t. and minimum sample period of 12 u.t.

In each of the three filters in this figure, the critical path is shown with dotted lines and addition and multiplication are assumed to take 1 and 2 u.t., respectively.

The two quantities W(U, V) and D(U, V) are used in this algorithm. The quantity W(U, V) is the minimum number of registers on any path from node U to node V and D(U, V) is the maximum computation time among all paths from U to V with weight W(U, V).

Formally,

$$W(U,V) = \min\{w(p) : U \xrightarrow{p} V\}$$

$$D(U,V) = \max\{t(p) : U \xrightarrow{p} V \text{ and } w(p) = W(U,V)\}.$$

The following algorithm can be used to compute W(U, V) and D(U, V):

- Let $M = t_{max} n$, where t_{max} is the maximum computation time of the nodes in G and n is the number of nodes in G.
- Form a new graph G′ which is the same as G except the edge weights are replaced by w′(e) = Mw(e) - t(U) for all edges $U \xrightarrow{e} V$.

- Solve the all-pairs shortest path problem on G′. Let S'_{UV} be the shortest path from U to V.
- If U ≠ V, then $W(U,V)=\left\lceil \frac{S'_{UV}}{M} \right\rceil$ and D(U, V) = MW(U, V) - S'_{UV} + t(V). If U = V, then W(U, V) = 0 and D(U, V) = t(U).

The value of $\lceil x \rceil$ is the ceiling of x, which is the smallest integer greater than or equal to x.

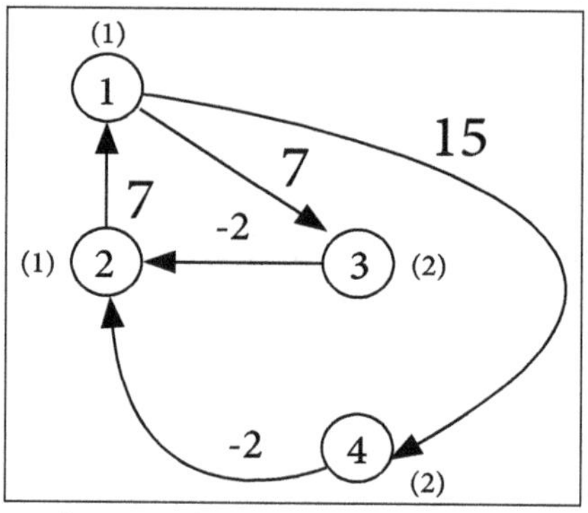

(8) The graph G′ used to compute W(U, V) and D(U,V).

To demonstrate this algorithm, the values of W(U, V) and D(U, V) are computed for the DFG. In the first step, $t_{max} = 2$ and n = 4, so M = 8. The new graph G′ is shown in figure (8). The solution to the all-pairs shortest path problem for G′ can be found using the Floyd-Warshall algorithm. This solution and the values of W(U, V) and D(U, V) are given in the below table.

The values of W(U, V) and D(U, V) are used to determine if there is a retiming solution that can achieve a desired clock period. Given a desired clock period c, there is a feasible retiming solution r such that $\phi(G_r) \le c$ if the following constraints hold.

- (Feasibility constraint) r(U) - r(V) ≤ w(e) for every edge $U \xrightarrow{e} V$ of G and,
- (Critical path constraint) r(U) - r(V) ≤ W(U, V) - 1 for all vertices U, V in G such that D(U, V) > c.

The feasibility constraint forces the number of delays on each edge in the retimed graph to be non-negative and the critical path constraint enforces φ(G) ≤ c.

Values of S'_{UV}, W(U, V) and D(U, V):

S'_{UV}	1	2	3	4
1	12	5	7	15
2	7	12	14	22

3	5	-2	12	20
4	5	-2	12	20

W(U, V)	1	2	3	4
1	0	1	1	2
2	1	0	2	3
3	1	0	0	3
4	1	0	2	0

D(U, V)	1	2	3	4
1	1	4	3	3
2	2	1	4	4
3	4	3	2	6
4	4	3	6	2

If D(U, V) > c, then W(U, V) + r(V) - r(U) ≥ 1 must hold for the critical path to have computation time less than or equal to c. This leads to the critical path constraint. For the DFG in figure (2a), if c is chosen to be 3, the inequalities r(U) - r(V) ≤ w(e) for every edge $U \xrightarrow{e} V$ are,

$$r(1) - r(3) \leq 1$$

$$r(1) - r(4) \leq 2$$

$$r(2) - r(1) \leq 1$$

$$r(3) - r(2) \leq 0$$

$$r(4) - r(2) \leq 0 \qquad \text{...(2)}$$

and the inequalities r(U) - r(V) ≤ W(U, V) - 1 for all vertices U, V in G such that D(U, V) > 3, found using the above table, are,

$$r(1) - r(2) \leq 0$$

$$r(2) - r(3) \leq 1$$

$$r(2) - r(4) \leq 2$$

$$r(3) - r(1) \leq 0$$

$$r(3) - r(4) \leq 2$$

$$r(4) - r(1) \leq 0$$

$$r(4) - r(3) \leq 1 \qquad \text{...(3)}$$

If there is a solution to the 12 inequalities in equation (2) and (3), then the solution is a feasible retiming solution such that the circuit can be clocked with period c = 3. Checking for a solution can be performed by constructing a constraint graph. The constraint graph for the 12 inequalities in equation (2) and (3) is shown in figure (9).

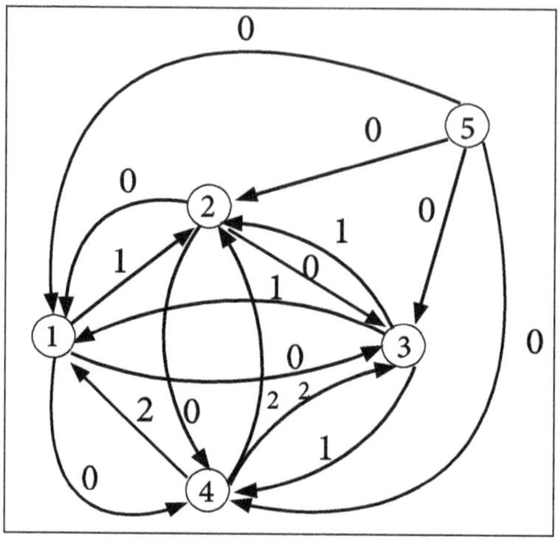

(9) The constraint graph.

Using either the Bellman-Ford algorithm or the Floyd-Warshall algorithm, it can be determined that the constraint graph in figure (9) contains no negative cycles, so the retiming solution for c = 3 is the solution to the single-source shortest path problem with the origin at the node 5, which is r(1) = r(2) = r(3) = r(4) = 0. This is intuitive since the DFG in figure (2a) already has φ(G) = 3, so no retiming needs to be performed to achieve a clock period of 3 u.t.

The fact that the 12 inequalities in equation (2) and (3) have a solution indicates that the minimum clock period for the circuit in figure (2a) is less than or equal to 3. It has been shown that the minimum clock period for the DFG is D(U, V) for some pair U, V. To find the minimum feasible clock period, we can attempt to solve the retiming equations for c equal to each unique value of D(U, V) and find the minimum value of c with a solution.

A clock period of c = 1 is not feasible because the nodes 3 and 4 have computation time greater than 1 u.t. A clock period of c = 3 is feasible, because the constraint graph in figure (9) has a solution, so in order to determine the minimum feasible clock period for the DFG in figure (2a), only the value of c = 2 must be checked.

For c = 2, the inequalities r(U) - r(V) ≤ w(e) for every edge $U \xrightarrow{e} V$ are given in equation (2) and the inequalities r(U) - r(V) ≤ W(U,V) - 1 for all vertices U, V in G such that D(U, V) > 2 are,

$$r(1) - r(2) \leq 0$$

$$r(1) - r(3) \leq 0$$

$r(1) - r(4) \leq 1$

$r(2) - r(3) \leq 1$

$r(2) - r(4) \leq 2$

$r(3) - r(1) \leq 0$

$r(3) - r(2) \leq -1$

$r(3) - r(4) \leq 2$

$r(4) - r(1) \leq 0$

$r(4) - r(2) \leq -1$

$r(4) - r(3) \leq 1$ ...(4)

The equations for a feasible retiming solution with clock period c = 2 are given in equation (2) and (4). Combining the equations in (2) and (4) that have identical left-hand sides results in the set of equations,

$r(1) - r(3) \leq 0$

$r(1) - r(4) \leq 1$

$r(2) - r(1) \leq 1$

$r(3) - r(2) \leq -1$

$r(4) - r(2) \leq -1$

$r(1) - r(2) \leq 0$

$r(2) - r(3) \leq 1$

$r(2) - r(4) \leq 2$

$r(3) - r(1) \leq 0$

$r(3) - r(4) \leq 2$

$r(4) - r(1) \leq 0$

$r(4) - r(3) \leq 1$ ...(5)

The constraint graph for the 12 inequalities in equation (5) is shown in figure (10). Using the Bellman-Ford algorithm or Floyd-Warshall algorithm, it can be determined that this graph contains no negative cycles, so the retiming solution for c = 2 is the solution to the single-source shortest path problem with the origin at the node 5, which is r(1) = -1, r(2) = 0, r(3) = -1 and r(4) = -1. This solution, which has clock period $\phi(G_r) = 2$, is the retimed version of figure (2a) and is shown in figure (2b).

The general approach to finding a retiming solution with the minimum clock period is to 1st compute W(U, V) and D(U, V) and then sort the values of D(U, V) and perform a binary search on these values to find the retiming solution with the minimum clock period. Computation of W(U, V) and D(U, V) can be done in $O(n^3)$ time using the Floyd-Warshall algorithm. The maximum number of distinct values of D(U,V) is n^2, so the binary search takes O(log n) steps.

Checking if one of these values is greater than or equal to the minimum clock period takes $O(n^3)$ time using either the Floyd-Warshall algorithm or the Bellman-Ford algorithm. Therefore, the time complexity required to find a retimed circuit with the minimum clock period is $O(n^3 \log n)$. Note that if the iteration bound is known, then only those values in D(U, V) that are greater than or equal to the iteration bound should be considered.

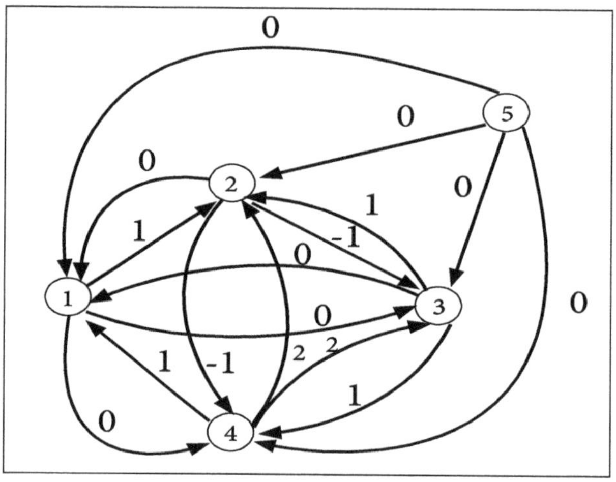

(10) The constraint graph.

Retiming for Register Minimization

If a node has several output edges carrying the same signal, the number of registers required to implement these edges is the maximum number of registers on any one of the edges. This is demonstrated in figure (11), where the naive implementation in figure (11a) uses 1 + 3 + 7 = 11 registers while the clever implementation in figure (11b) uses max(1, 3, 7) = 7 registers.

Using this concept, the number of registers required to implement the output edges of the node V in the retimed graph is,

$$R_v = \max_{v \xrightarrow{r} ?} \{ wr(e) \}$$

And the total register cost in the retimed circuit is, $COST = \sum R_v$.

The formulation of retiming to minimize the number of registers under the constraint that the clock period is not greater than c is: Minimize $COST = \sum R_v$ subject to,

- (Fan-out constraint) $R_v \geq w_r(e)$ for all V and all edges $V \xrightarrow{e} ?$.
- (Feasibility constraint) r(U) - r(V) ≤ w(e) for every edge $U \xrightarrow{e} V$.
- (Clock period constraint) r(U) - r(V) ≤ W(U, V) - 1 for all vertices U, V such that D(U, V) > c.

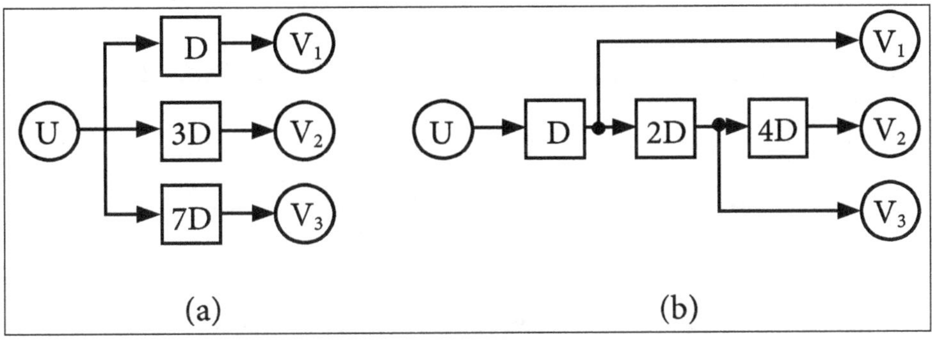

(11) Fan-out implementation using 1 + 3 + 7 = 11 registers and Fan-out implementation using max(1, 3, 7) = 7 registers.

The fan out constraint simply makes sure that $R_V = \max_{V \xrightarrow{e} ?} \{w_r(e)\}$, along with descriptions of the feasibility constraint and clock period constraint.

While this formulation of retiming indeed minimizes the number of registers under a clock period constraint, it is not in a form that can be solved using Linear Programming (LP) techniques, because some solutions may not be integers. To get the formulation in such a form, a "gadget" is used to represent nodes with multiple output edges. This gadget is shown in figure (12). Figure (12a) shows a fan out node with k output edges.

The gadget in figure (12b) is used to model the fan out node. Each of the k edges e_i, $1 \leq i \leq k$, has an associated weight $w(e\ )$ which is known from the DFG. The node $\hat{U}$ is a dummy node with zero computation time ($t(\hat{U}) = 0$) and the edges e_i, $1 \leq i \leq k$, are dummy edges introduced so the retiming for register minimization problem can be modeled as a linear programming problem.

The weight of the edge $\hat{e}_i$, is defined to be $w(\hat{e}_i) = w_{max} - w(e_i)$, where $w_{max} = \max_1 \leq i \leq k\, w(e_i)$. In addition to its weight, each edge in this model also has a breadth β associated with it. This breadth of an edge is simply a number used so that the gadget in figure (12b) properly models the memory required by the edges e_i, $1 \leq i \leq k$, in the retimed DFG. The breadth of each of the edges e_i and $\hat{e}_i$ for $1 \leq i \leq k$ is β = 1/k, as shown in figure (12b).

Using the fan out model in figure (12), the retiming formulation is: Minimize $COST = \Sigma_{(e)} \beta(e) w_r(e)$ subject to,

- (Feasibility constraint) r(U) - r(V) ≤ w(e) for every edge $U \xrightarrow{e} V$.
- (Clock period constraint) r(U) - r(V) ≤ W(U, V) -1 for all vertices U, V such that D(U, V) > c.

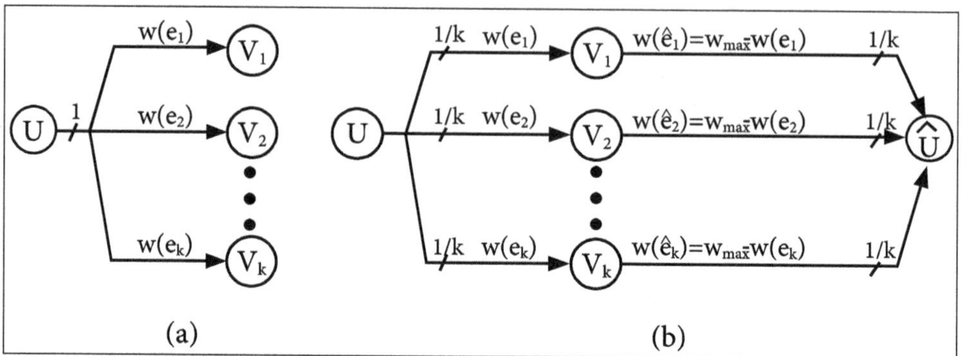

(12) A node U with k output edges and A gadget used to model the node U.

The expression for COST can be rewritten as,

$$\begin{aligned}
COST &= \sum_{e} \beta(e) w_r(e) \\
&= \sum_{e} \beta(e)\big(w(e)+r(V)-r(U)\big) \\
&= \sum_{e} \beta(e) w(e) + \sum_{e} \beta(e)\big(r(V)-r(U)\big) \\
&= K + \sum_{e} \beta(e)\big(r(V)-r(U)\big) \\
&= K + \sum_{V} r(V)\left(\sum_{? \xrightarrow{e} V} \beta(e) - \sum_{V \xrightarrow{e} ?} \beta(e) \right)
\end{aligned}$$

Since K is a constant, the formulation can be written as Minimize COST′ = $\Sigma_V r(V)\big(\Sigma_{? \xrightarrow{e} V} \beta(e) - \Sigma_{V \xrightarrow{e} ?} \beta(e)\big)$ subject to,

- (Feasibility constraint) r(U) - r(V) ≤ w(e) for every edge $U \xrightarrow{e} V$.
- (Clock period constraint) r(U) - r(V) ≤ W(U, V) - 1 for all vertices U, V such that D(U, V) > c.

To demonstrate this, the DFG in figure (2a) is retimed so the number of registers is minimized while maintaining a clock period of $\phi(G_r) \leq 2$. The 1st step is to redraw this DFG as shown in figure (13a) using the fan out model for node 1, because this node has more than one output edge. The 2nd step is to compute W(U, V) and D(U,V) for the DFG in figure (13a).

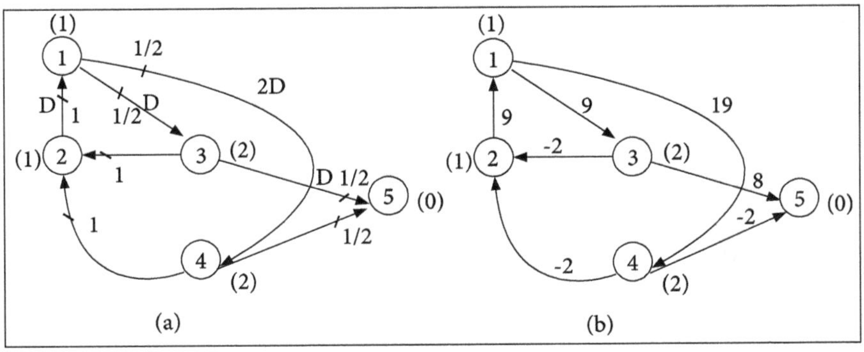

(13) The DFG in figure (2a) redrawn using the fan out gadget in figure (12) and The graph G′ used to compute W(U,V) and D(U, V).

Values of S'_{UV}, W(U,V) and D(U, V) for the DFG in figure (13a):

S'_{UV}	1	2	3	4	5
1	16	7	9	19	17
2	9	16	18	28	26
3	7	-2	16	26	8
4	7	-2	16	26	-2
5	-	-	-	-	-

W(U, V)	1	2	3	4	5
1	0	1	1	2	2
2	1	0	2	3	3
3	1	0	0	3	1
4	1	0	2	0	0
5	-	-	-	-	0

D(U, V)	1	2	3	4	5
1	1	4	3	3	3
2	2	1	4	4	4
3	4	3	2	6	2
4	4	3	6	2	2
5	-	-	-	-	0

The LP formulation of the problem is: Minimize

$$COST' = r(1)\,(1 - 1) + r(2)\,(2 - 1) + r(3)\,(1/2 - 3/2) + r(4)\,(1/2 - 3/2) + r(5)\,(1 - 0)$$

$$= r(2) - r(3) - r(4) + r(5)$$

Subject to the feasibility constraints,

$r(1) - r(3) \leq 1$

$r(1) - r(4) \leq 2$

$r(2) - r(1) \leq 1$

$r(3) - r(2) \leq 0$

$r(3) - r(5) \leq 1$

$r(4) - r(2) \leq 0$

$r(4) - r(5) \leq 0$

And the clock period constraints,

$r(1) - r(2) \leq 0$

$r(1) - r(3) \leq 0$

$r(1) - r(4) \leq 1$

$r(1) - r(5) \leq 1$

$r(2) - r(3) \leq 1$

$r(2) - r(4) \leq 2$

$r(2) - r(5) \leq 2$

$r(3) - r(1) \leq 0$

$r(3) - r(2) \leq -1$

$r(3) - r(4) \leq 2$

$r(4) - r(1) \leq 0$

$r(4) - r(2) \leq -1$

$r(4) - r(3) \leq 1$

This linear programming problem is in the form of a minimum-cost flow problem, for which algorithms exist; however, these algorithms can be quite involved. To solve this problem, a linear programming software package such as GAMS can be used. Using GAMS, the solution to this problem is $r(1) = 1$, $r(2) = 2$, $r(3) = 1$, $r(4) = 0$ and $r(5) = 0$ and the retimed graph, which uses proper register sharing between the 2 output edges of the node 1, is shown in figure (14).

It remains to show that the fan out gadget in figure (12b) properly models the amount of memory used by a node with multiple output edges. The goal is for the sum of the weights of all of the edges e_i and $\hat{e}_i$, in the fan out gadget to equal the maximum of the weights on the edges e_i in the gadget. In other words, the goal is to show that,

$$\sum_{i=1}^{k}\left(w_r(e_i)\beta(e_i)+w_r(\hat{e}_i)\beta(\hat{e}_i)\right)=w_{r,max}$$

Where,

$$w_{r,\,max}=\max_{1\le i\le k} w_r(e_i)$$

This is shown using the following three steps:

Step 1: Assume that $w_r(e_j)=w_{r,\,max}$. Show that $w_r(e_j)=w_{max}+r(\hat{U})-r(U)$.

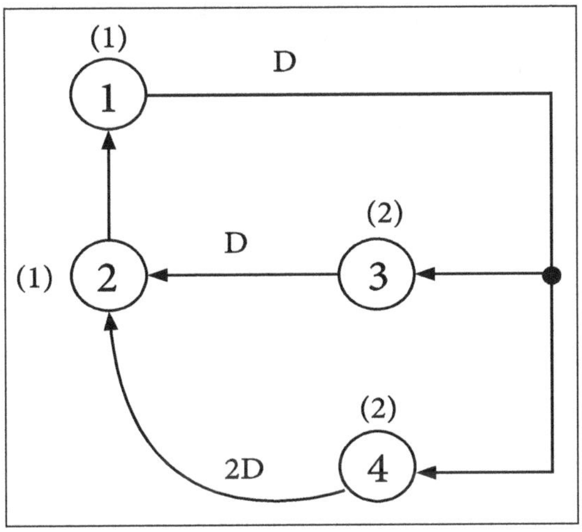

(14) The result of retiming the DFG in figure (2a) to minimize the number of registers while maintaining a clock period of 2 u.t.

Step 2: Show that,

$$\sum_{i=1}^{k}\left(w_r(e_i)\beta(e_i)+w_r(\hat{e}_i)\beta(\hat{e}_i)\right)=w_{max}+r(\hat{U})-r(U)$$

Step 3: Combining steps 1 and 2 results in,

$$\sum_{i=1}^{k}\left(w_r(e_i)\beta(e_i)+w_r(\hat{e}_i)\beta(\hat{e}_i)\right)=w_{max}+r(\hat{U})-r(U)=w_r(e_j)=w_{r,max}$$

While step 3 is trivial, steps 1 and 2 are not trivial.

Details of Step 1: Assume that $w_r(e_j)=w_{r,\,max}$. The weight of each $p_i=U\xrightarrow{e}V_i\xrightarrow{e}\hat{U}$ in the original DFG is,

$$w(p_i)=w(e_i)+\left(w_{max}-w(e_i)\right)$$

$$= w_{max}$$

The weight of each path p_i in the retimed DFG is given by,

$$
\begin{aligned}
w_r(p_i) &= w_r(e_i) + w_r(\hat{e}_i) \\
&= \left(w(e_i) + r(V_i) - r(U)\right) + \left(w_{max} - w(e_i) + r(\hat{U}) - r(V_i)\right) \\
&= w_{max} + r(\hat{U}) - r(U).
\end{aligned}
$$

Solving to minimize COST causes $r(\hat{U})$ to be as small as possible while still forcing $w_r(\hat{e}_i) \geq 0$ for $1 \leq i \leq k$. Therefore, $w_r(\hat{e}_i) = 0$ for at least 1 edge $\hat{e}_i$. Since the value of $w_r(p_i) = w_r(e_i) + w_r(\hat{e}_i) = w_{max} + r(\hat{U}) - r(U)$ is the same for $1 \leq i \leq k$, the edge e_j with $w_r(e_j) = w_{r,\,max}$ is in the path p_j which has $w_r(e_j) = 0$. Thus, $w_r(e_j) = w_r(p_j) = w_{max} + r(\hat{U}) - r(U)$.

Details of Step 2: The cost of the edges in the fan out model is,

$$
\begin{aligned}
COST &= \sum_{i=1}^{k}\left(w_r(e_i)\beta(e_i) + w_r(\hat{e}_i)\beta(\hat{e}_i)\right) \\
&= \frac{1}{k}\left(\sum_{i=1}^{k}\left(w_r(e_i) + w_r(\hat{e}_i)\right)\right) \\
&= \frac{1}{k}(k)\left(w_{max} + r(\hat{U}) - r(U)\right)
\end{aligned}
$$

1.6 Unfolding: An Algorithm for Unfolding

Unfolding is a transformation technique that can be applied to a DSP equation to create a new equation describing more than one iteration of the original equation. More specifically, unfolding a DSP equation by the unfolding factor J creates a new equation that describes J consecutive iterations of the original equation. Unfolding is also referred to as loop unrolling and has been used in compiler theory.

For example, consider the below equation,

$$y(n) = ay(n-9) + x(n) \qquad \text{...(1)}$$

For n = 0 to ∞, which is shown in figure (1a).

Replacing the index n with 2k results in y(2k) = ay(2k - 9) + x(2k) for k = 0 to ∞. Similarly, replacing the index n with 2k + 1 results in y(2k + 1) = ay(2k - 8) + x(2k + 1) for k = 0 to ∞. Together, the following two equations describe the same as equation (1).

$$y(2k) = ay(2k-9) + x(2k)$$

$$y(2k+1) = ay(2k-8) + x(2k+1) \qquad ...(2)$$

Equation (2) describes two consecutive iterations of equation (1). For example, for k = 7, the equations in (2) are y(14) = ay(5) + x(14) and y(15) = ay(6) + x(15). These equations are the same as those described in equation (1) for the two consecutive iterations n = 14 and n = 15. Equation (2) describes a 2-unfolded version of equation (1).

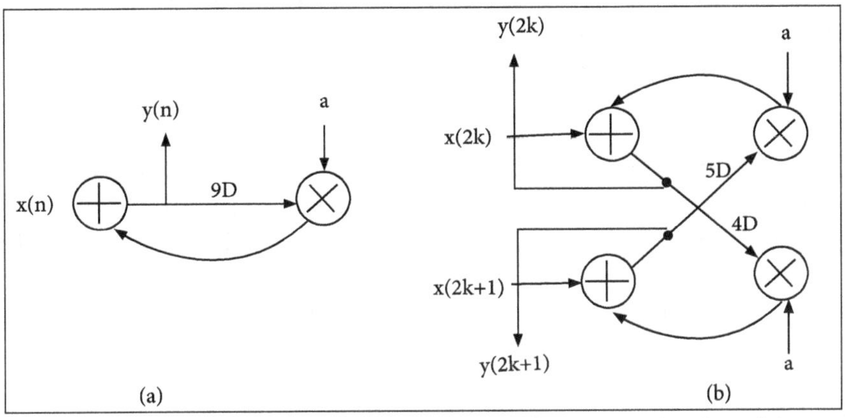

(1) The original DFG of DSP equation (1) and the 2-unfolded DSP equation describing equation (2).

In unfolded systems, each delay is J-slow. This means that if the input to a delay element is the signal x(kJ + m), then the output of the delay element is x((k - 1)J + m) = x(kJ + m - J). The 2-unfolded equation corresponding to equation (2) is shown in figure (1b). Using the concept of J-slow delays, figure (1b) describes the computations, y(2k) = ay(2(k - 5) + 1) + x(2k) and y(2k + 1) = ay(2(k - 4) + 0) + x(2k + 1), as desired.

The 2-unfolded equations in (2) are obtained by substituting n = 2k and n = 2k + 1 in equation (1). While this technique results in a correct 2-unfolded equation, it can be tedious to write the original and J-unfolded equations first and then draw the corresponding unfolded Data Flow Graph, especially for larger values of J.

Unfolding has applications in designing high speed and low power VLSI architectures. One application is to unfold the equation to reveal hidden concurrencies, so that the equation can be scheduled to a smaller iteration period, thus, increasing the throughput of the implementation.

Another application is to design a parallel architecture at the word level and bit level. At the word level, these architectures can be used to design word parallel architectures from word serial architectures to increase the throughput or decrease the power consumption of the implementation.

At the bit level, unfolding can be used to design a bit parallel and digit serial architectures from bit serial architectures to achieve the same objectives as word level unfolding. Here, we always assume that the node cannot be split into two sub-nodes with smaller computation times.

An Algorithm for Unfolding

The operation $\lfloor x \rfloor$ is the floor of x, which is the largest integer less than or equal to x. For example, $\lfloor 11/4 \rfloor = 2$ and $\lfloor 25/8 \rfloor = 8$. The operation a%b is the remainder after dividing a by b, where a and b are integers. For example, 11%4 = 3 and 25%3 = 1.

For each node U in the original DFG, there are J nodes with the same function as U in the J-unfolded DFG. For example, for the adder in figure (1a), there are two adders in the 2-unfolded DFG in figure (1b). Similarly, there are two multipliers in figure (1b) corresponding to the multiplier in figure (1a).

For each edge in the original DFG, there are J edges in the J-unfolded DFG. This can be observed in figure (1), where the edge from the adder to the multiplier in the original DFG results in two edges from the adder to the multiplier in the unfolded DFG.

DFG of the J-unfolded equation contains J times as many nodes and edges as the original DFG, the following unfolding algorithm can be used to construct a J-unfolded DFG.

Unfolding Algorithm

- For each node U in the original DFG, draw the J nodes U_0, U_1, ..., U_{J-1}.
- For each edge U → V with w delays in the original DFG, draw the J edges $U_i \rightarrow V_{(i+W)\%J}$ with $\left\lfloor \frac{i+w}{J} \right\rfloor$ delays for i = 0, 1, ..., J - 1.

To demonstrate the unfolding algorithm, a DFG is constructed for the DSP equation as in figure (1a) and this DFG is unfolded with unfolding factor 2 to obtain the DSP equation as in figure (1b). A DFG representing the DSP equation is shown in figure (2a).

In this DFG, the nodes A and B represent the input and output, respectively and the nodes C and D represent addition and multiplication by a, respectively. To unfold this DFG by unfolding factor 2, the 8 nodes A_i, B_i, C_i and D_i, for i = 0, 1, are first drawn according to the 1st step of the unfolding algorithm. After these nodes have been drawn, the 2nd step of the unfolding algorithm needs to be performed. For an edge $U \rightarrow V$ with no delays, this step reduces to drawing the J edges $U_i \rightarrow V_i$ with no delays for i = 0, 1, ..., J - 1.

For example, the edge D → C with no delays in figure (2a) results in two edges $D_0 \rightarrow C_0$ and $D_1 \rightarrow C_1$ with no delays in the 2-unfolded DFG in figure (2b). For the edge C → D with w = 9 delays in figure (2a), we draw the edges $C_0 \rightarrow D_{(9+0)\%2}$ with $\left\lfloor \frac{9+0}{2} \right\rfloor$ delays and $C_1 \rightarrow D_{(9+1)\%2}$ with $\left\lfloor \frac{9+1}{2} \right\rfloor$ delays, which correspond to the edges $C_0 \rightarrow D_1$ with four delays and $C_1 \rightarrow D_0$ with five delays, respectively, in figure (2b).

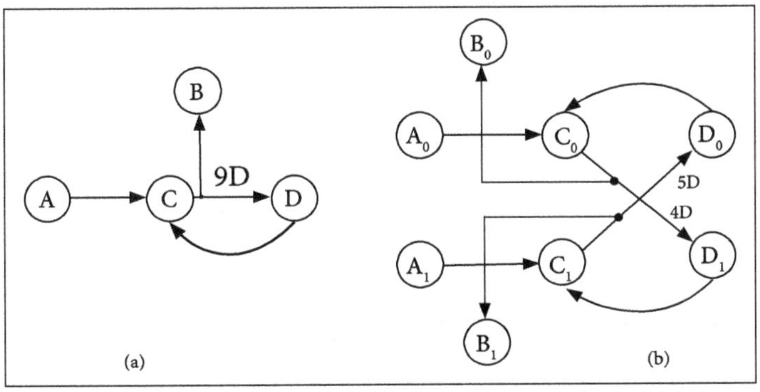

(2) A DFG corresponding to the DSP equation in figure (1a) and The 2-unfolded DFG corresponding to the 2-unfolded DSP equation in figure (1b).

Referring to figure (2), the nodes C_0 and C_1 in the 2-unfolded DFG represent addition, because the node C in the original DFG represents addition. Similarly, the nodes D_0 and D_1 in the 2-unfolded DFG represent multiplication by a, because the node D in the original DFG represents multiplication by a. The node A in the original DFG represents the input x(n). The kth iteration of the node A_i in the unfolded DFG executes the $(Jk+i)^{th}$ iteration of the node A in the original DFG for i = 0,1, ..., J - 1 and k = 0 to ∞, so the node A_0 corresponds to the input samples x(2k + 0) and the node A_1 corresponds to the input samples x(2k + 1).

Similarly, the node B_0 corresponds to the output samples y(2k + 0) and the node B_1 corresponds to the output samples y(2k + 1). Hence, the 2-unfolded DFG in figure (2b) corresponds to the 2-unfolded DSP equation in figure (1b).

Two examples of unfolding are shown in figure (3) and (4). The DFG in figure (3a) is unfolded using unfolding factor 4 to get the DFG in figure (3b) and the DFG in figure (4a) is unfolded using unfolding factor 3 to get the DFG in figure (4b).

Unfolding of an edge with w delays in the original DFG produces J - w edges with no delays and w edges with 1 delay in J-unfolded DFG when w < J.

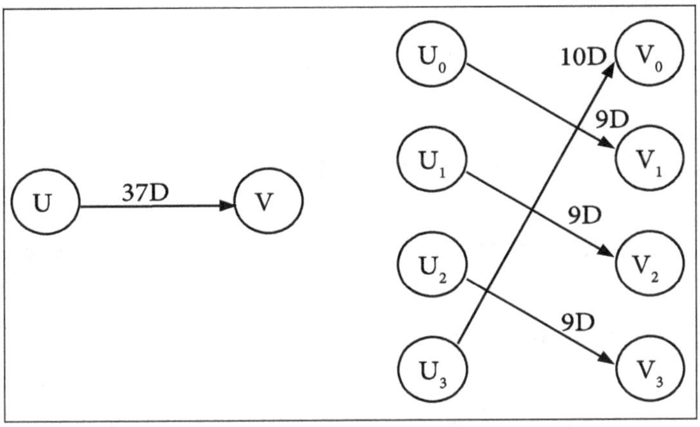

(3) The original DFG and The 4-unfolded DFG.

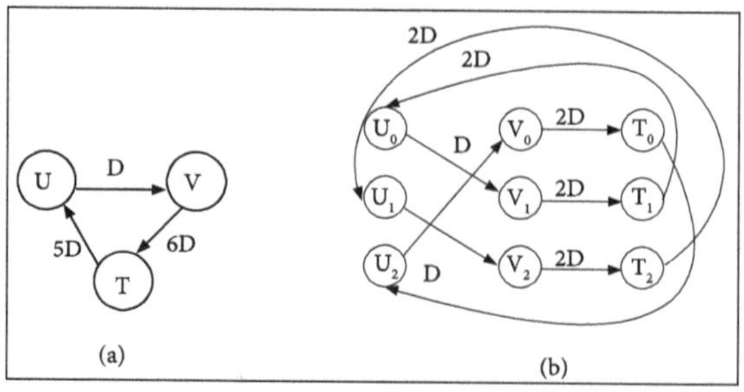

(4) The original DFG and The 3-unfolded DFG.

An important property of unfolding is that unfolding preserves the precedence constraints of a DSP equation. The e edges in the original DFG explicitly show the precedence constraints for one iteration of the original equation and the Je edges in the J-unfolded DFG explicitly show the precedence constraints for J iterations of the equation.

To show that the unfolding algorithm creates an unfolded DFG with the same precedence constraints as the original DFG, we need to show that the edge $U_i \rightarrow V_{(i+W)\%J}$ with $\left\lfloor \frac{i+w}{J} \right\rfloor$ delays in the unfolded DFG corresponds to the edge $U \rightarrow V$ with w delays in the original DFG.

The unfolding transformation is based on the fact that, the kth iteration of the node U_i in the J-unfolded DFG executes the $(Jk+i)^{th}$ iteration of the node U in the original DFG. Due to the $\left\lfloor \frac{i+w}{j} \right\rfloor$ delays on the edge $U_i \rightarrow V_{(i+w)\%J}$, the output of the kth iteration of the node U_i is consumed by the $\left(k+\left\lfloor \frac{i+w}{j} \right\rfloor\right)^{th}$ iteration of the node $V_{(i+w)\%J}$ in the unfolded DFG.

The kth iteration of U_i corresponds to the $(Jk+i)^{th}$ iteration of the node U and the $\left(k+\left\lfloor \frac{i+w}{j} \right\rfloor\right)^{th}$ iteration of the node $V_{(i+w)\%J}$ corresponds to the $\left(J\left(k+\left\lfloor \frac{i+w}{J} \right\rfloor\right)+(i+w)\%J\right)^{th}$ iteration of the node V. Therefore, in the original DFG, the output of the $(Jk+i)^{th}$ iteration of the node U is consumed by the $\left(J\left(k+\left\lfloor \frac{i+w}{J} \right\rfloor\right)+(i+w)\%J\right)^{th}$ iteration of the node V, so the number of delays on the edge $U \rightarrow V$ is,

$$J\left(k+\left\lfloor \frac{i+w}{J} \right\rfloor\right)+(i+w)\%J-(Jk+i)$$

Which can be simplified as,

$$\left(J\left\lfloor\frac{i+w}{J}\right\rfloor+(i+w)\%J\right)-i$$

The value,

$$J\left\lfloor\frac{i+w}{J}\right\rfloor+(i+w)\%J$$

is simply $i + w$, so the number of delays on the edge $U \rightarrow V$ is $(i + w)-i=w$.

1.6.1 Properties of Unfolding

These properties are useful when applications of unfolding are considered.

Property-1

Unfolding preserves the number of delays in a DFG.

This property is based on the fact that, the sum of the delays on the J unfolded edges $U_i \rightarrow V_{(i+w)\%J}$, where, i = 0, 1, ..., J - 1, is same as the number of delays on the edge U → V in the original DFG.

Mathematically, this can be stated as,

$$\left\lfloor\frac{w}{J}\right\rfloor+\left\lfloor\frac{w+1}{J}\right\rfloor+\cdots+\left\lfloor\frac{w+J-1}{J}\right\rfloor=w$$

This property can be observed in figures (2), (3) and (4).

Let l be a loop with w_l delays in the original DFG and let A be a node in l. The loop l can be denoted as the path $A\xrightarrow{l}A$ with w_l delays. If the loop l is traversed p times (p ≥ 1), this results in the path $A\xrightarrow{l}A\xrightarrow{l}\ldots\xrightarrow{l}A$ with pw_l delays.

The corresponding unfolded path starting at the node A_i, 0 ≤ i ≤ J - 1, in the J-unfolded DFG is $A_i \rightarrow A_{(i+pw_l)\%J}$ and this path forms a loop in the unfolded DFG if $i=(i+pw_l)\%J$. We would like to find the minimum value of p (i.e.) the minimum number of times that the loop l is traversed, such that the path $A_i \rightarrow A_{(i+pw_l)\%J}$ is a loop in the unfolded DFG.

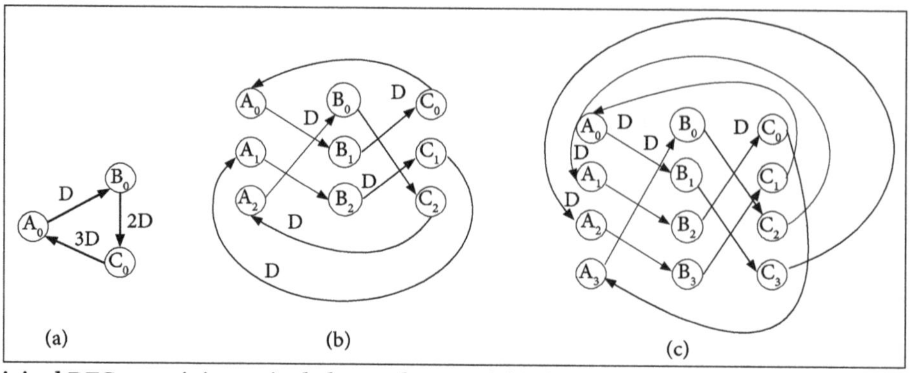

(5) The original DFG containing a single loop, The 3-unfolded DFG containing 3 loops and The 4-unfolded DFG containing 2 loops.

For example, the DFG in figure (5a) contains a single loop $l = A \rightarrow B \rightarrow C \rightarrow A$ with $w_l = 6$ delays. The 3-unfolded and 4-unfolded DFGs are shown in figures (5b) and (5c), respectively. For J = 3, the equation $i = (i + pw_l)\%J$ becomes i = (i + 6p)%3, which holds for i = 0, 1, 2 for p = 1. Therefore, unfolding p = 1 instance of the loop results in a loop in the 3-unfolded DFG.

Starting at A_0, unfolding the loop l results in the loop $A_0 \rightarrow B_1 \rightarrow C_0 \rightarrow A_0$. Starting at A_1, unfolding the loop l results in the loop $A_1 \rightarrow B_2 \rightarrow C_1 \rightarrow A_1$. Starting at A_2, unfolding the loop l results in the loop $A_2 \rightarrow B_0 \rightarrow C_2 \rightarrow A_2$. The 3-unfolded DFG consists of these 3 loops. For J = 4, the equation $i = (i + pw_l)\%J$ becomes i = (i + 6p)%4, which does not hold for p = 1, but holds for i = 0, 1, 2, 3 for p = 2. Therefore, unfolding p = 2 instances of the loop results in a loop in the 4-unfolded DFG.

Starting at A_0, unfolding the loop $l' = A \rightarrow B \rightarrow C \rightarrow A \rightarrow B \rightarrow C \rightarrow A$ results in the loop $A_0 \rightarrow B_1 \rightarrow C_3 \rightarrow A_2 \rightarrow B_3 \rightarrow C_1 \rightarrow A_0$. Stating at A_1, unfolding the loop l' results in the loop $A_1 \rightarrow B_2 \rightarrow C_0 \rightarrow A_3 \rightarrow B_0 \rightarrow C_2 \rightarrow A_1$. The 4-unfolded DFG consists of these 2 loops.

When a loop is unfolded, the number of resulting unfolded loops can be determined without going through the process of unfolding. This number can be determined by finding the minimum value of p, such that the path $A_{(i+pwl)\%J}$ forms a loop, (i.e.) finding the minimum value of p, such that $i = (i + pw_l)\%J$.

The following two lemmas are used to show this:

Lemma (1): $i = (i + pw_l)\%J \Leftrightarrow pw_l = q_J$ for an integer q.

Lemma (2): The smallest positive integer p that satisfies $pw_l = q_J$ is $J/\gcd(w_l, J)$.

Proof:

The expression $pw_l = q_J$ can be written as $\left(\frac{a}{w_l}\right)w_l = \left(\frac{a}{J}\right)J$. The values $p = (a / w_l)$ and $q = (a/J)$ must be integers and a is to be as small as possible, because $p = (a / w_l)$ is to be as small as possible, so $a = \text{lcm}\{w_l, J\}$. Using the fact that, $\text{lcm}\{w_l, J\}\ \text{gcd}\{w_l, J\} = w_l J$, the minimum value of p is $p = \frac{\text{lcm}\{w_l, J\}}{w_l} = \frac{J}{\text{gcd}\{w_l, J\}}$.

From Lemma (1), the path $A_i \rightarrow A_{(i+nw)\%J}$ is a loop, if and only if $pw_l = q_J$ holds. The minimum value of p that satisfies this is equal to the number of times the loop l is traversed to form an unfolded loop and Lemma (2) states that this number is $J/\text{gcd}(w_l, J)$, where gcd(a, b) represents the greatest common divisor of a and b. Therefore, an unfolded loop contains $J/\text{gcd}(w_l, J)$ copies of each node in l.

The unfolded DFG contains a total of J copies of each node, so there must be $J/(J/\text{gcd}(w_l, J)) = \text{gcd}(w_l, J)$ unfolded loops that result from unfolding l. The unfolded DFG contains w_l delays and each of the $\text{gcd}(w_l, J)$ unfolded loops contain $w_l/\text{gcd}(w_l, J)$ delays.

Property-2

J-unfolding of a loop l with w_l delays in the original DFG leads to $\text{gcd}(w_l, J)$ loops in the unfolded DFG and each of these $\text{gcd}(w_l, J)$ loops contain $w_l/\text{gcd}(w_l, J)$ delays and $J/\text{gcd}(w_l, J)$ copies of each node that appears in l.

The following property describes how the iteration bound is affected by unfolding:

Property-3

Unfolding a DFG with iteration bound T_∞ results in a J-unfolded DFG with iteration bound JT_∞.

Since the J-unfolded DFG represents J iterations of the original DFG, the iteration bound of the J-unfolded DFG should be J times the iteration bound of the original DFG. Mathematically, the iteration bound of the original DFG is, $T_\infty = \max_l \left\{\frac{t_l}{w_l}\right\}$

and property (2) can be used to show that the iteration bound of the unfolded DFG is,

$$T'_{\infty} = \max_{l}\left\{\frac{\left(J/\gcd(w_l,J)\right)t_l}{w_l/\gcd(w_l,J)}\right\} = J\max_{l}\left\{\frac{t_l}{w_l}\right\} = JT_{\infty}$$

1.6.2 Critical Path, Unfolding and Retiming

Property-1

Consider a path with w delays in the original DFG. J-unfolding of this path leads to (J - w) paths with no delays and w paths with 1 delay each, when w < J.

Using this property, we can retime the original DFG, such that the J-unfolded version of the retimed DFG will meet a specified critical path computation time, c. The critical path of the unfolded DFG can be c, if there exists a path in the original DFG with computation time c and less than J delay elements.

This leads to the critical path constraint for retiming for reducing the critical path in the unfolded version of the retimed DFG. If D(U, V) ≥ c, then $W_r(U, V) = W(U, V) + r(V) - r(U) \geq J$ or $r(U) - r(V) \leq W(U, V) - J$. The usual feasibility constraint, $w(e) + r(V) - r(U) \leq 0$, should be used with this critical path constraint.

Lemma: Any feasible clock cycle period that can be obtained by retiming the J-unfolded DFG, G_J, can be achieved by retiming the original DFG, G, directly and then unfolding it by unfolding factor J.

Proof:

Let r' be a legal retiming for the unfolded DFG, G_J, which leads to critical path c. Let r be the retiming for G defined as,

$$r(U) = \sum_{i=0}^{J-1} r^{l}(U_i)$$

We prove that r is a feasible retiming on the original DFG, such that the J-unfolded version of the DFG, G_r, will have a critical path c. Consider an edge U → V with w delays in G. Since r' is a legal retiming on the unfolded DFG, it can be written as,

$$r'(U_i) - r'\left(V_{(i+w)\%J}\right) \leq \left\lfloor \frac{i+w}{J} \right\rfloor$$

Summing the above inequalities for i = 0 to J - 1, we obtain r(U) - r(V) ≤ w, which proves that r is a feasible retiming. Now consider the proof, that the critical path in the

J-unfolded DFG is c. Since r' satisfies the critical path constraint for the unfolded DFG, (i.e.),

$$r'(U_i) - r'\left(V_{(i+w)\%J}\right) \le \left\lfloor \frac{i+w}{J} \right\rfloor - 1$$

If $D\left(U_i \to V_{(i+w)\%J}\right) > c$, summing this for all i ranging from 0 to J - 1 results in,

$$r(U) - r(V) \le W(U, V) - J$$

Which is the desired inequality.

1.6.3 Applications of Unfolding

Sample Period Reduction

The iteration bound, T_∞, is the lower bound on the iteration period of a recursive DSP equation. An implementation of the DSP equation can never achieve an iteration period less than the iteration bound, even if infinite processors are available. In some cases, the DSP equation cannot be implemented with the iteration period equal to the iteration bound without the use of unfolding.

The 1st case where the iteration period cannot be made equal to the iteration bound is when there is a node in the DFG that has computation time greater than T_∞. In this case, even retiming cannot be used to reduce the computation time of the critical path of the DFG to T_∞. For example, the DFG in figure (6a) has iteration bound $T_\infty = 3$, but the nodes S and T each require computation times of 4 u.t., so the minimum sample period after retiming is 4 u.t.

This DFG can be unfolded with unfolding factor 2 to get the DFG in figure (6b). This unfolded DFG has iteration bound $T_\infty = 6$ and its critical path is 6 u.t. The unfolded DFG performs two iterations of the original DSP equation in 6 u.t., so the sample period of the unfolded DFG is 6/2 = 3 u.t., which is the same as the iteration bound of the original DFG.

The original DFG in figure (6a) cannot achieve a sample period equal to the iteration bound, because a node computation time is larger than the iteration bound, but the unfolded DFG in figure (6b) can have a sample period equal to the iteration bound of the original DFG. In general, if the computation time of node U, t_U, is greater than the iteration bound, T_∞, then $\lceil t_U / T_\infty \rceil$-unfolding should be used, where $\lceil x \rceil$ is the ceiling of x, which is the smallest integer greater than or equal to x. In the above example, $t_U = 4$ and $T_\infty = 3$, so $\lceil 4/3 \rceil$-unfolding (i.e.) 2-unfolding are used.

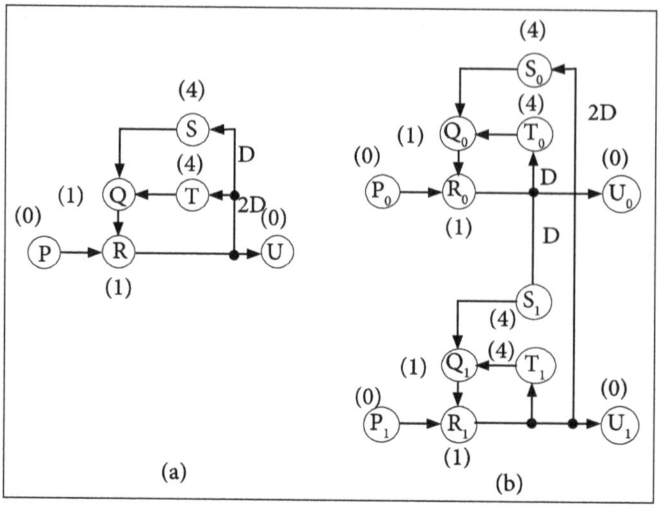

(6) The original DFG and The 2-unfolded DFG.

In figure (6a), the iteration bound, $T_\infty = 6$ and minimum sample period is 6/2 = 3 and in figure (6b), it2eration bound $T_\infty = 3$ and minimum sample period is 4.

The second case where the iteration period cannot be made equal to the iteration bound is when the iteration bound is not an integer. A simple example of this is shown in figure (7a), where the DFG has iteration bound $T_\infty = 4/3$, however, even retiming cannot be used to achieve a critical path of less than 2 u.t. This DFG can be unfolded with unfolding factor 3 to get the DFG in figure (7b). This unfolded DFG has iteration bound $T_\infty = 4$ and its critical path is 4 u.t. The unfolded DFG performs 3 iterations of the original DSP equation in 4 u.t., so the sample period of the unfolded DFG is 4/3 u.t., which is the same as the iteration bound of the original DFG.

The original DFG in figure (7a) cannot achieve a sample period equal to the iteration bound, because the iteration bound is not an integer, but the unfolded DFG in figure (7b) can have a sample period equal to the iteration bound of the original DFG. In general, if a critical loop bound is of the form t_l / w_l, where t_l and w_l are mutually co-prime, then w_l -unfolding should be used. For example, if $t_l = 60$ and $w_l = 45$, then t_l / w_l should be written as $t_l / w_l = 4/3$ and 3-unfolding should be used.

Situations may also occur where the longest node computation time is larger than the iteration bound, T_∞ and T_∞ is not an integer. In this case, the minimum unfolding factor that allows the iteration period to equal the iteration bound is the minimum value of J such that JT_∞ is an integer and is greater than or equal to the longest node computation time. For example, if $T_\infty = 4/3$ and the longest node computation time is 6, then the minimum unfolding factor that allows the iteration period to equal the iteration bound is J = 6, because this is the minimum value of J such that J(4/3) is an integer and J(4/3) ≥ 6.

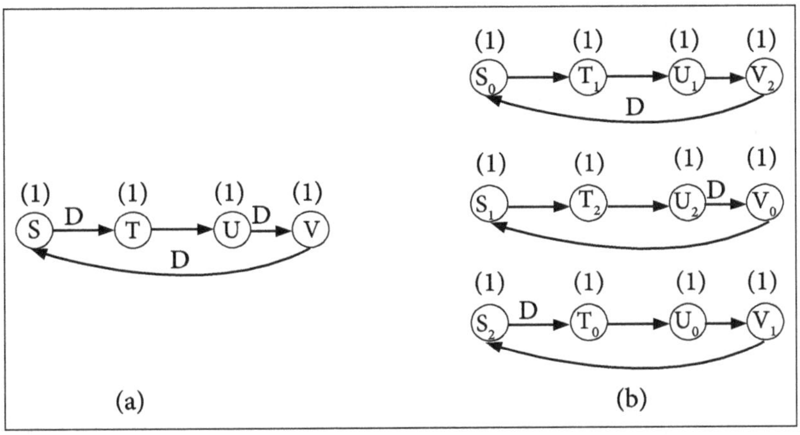

(7) The original DFG and The 3-unfolded DFG.

In figure (7a), the iteration bound $T_\infty = 4/3$ and minimum sample period is 2 and in figure (7b) the iteration bound $T_\infty = 4$ and minimum sample period is 4/3.

Any DFG with 1 delay in each loop is referred to as a perfect-rate DFG. A perfect-rate DFG can always be scheduled such that the iteration period is equal to the iteration bound of the DFG. If a DFG is not perfect rate, then it can always be unfolded, so the unfolded DFG is perfect rate. Therefore, if $\gcd(w_l, J) = w_l$, the unfolded loops contain 1 delay each and the unfolded DFG is perfect rate.

For $\gcd(w_l, J) = w_l$ to hold for all loops l in the original DFG, J must be a multiple of w_l for each loop l. Choosing J to be equal to the least common multiple of the number of loop delays guarantees that J is a multiple of w_l for each loop l, which guarantees that the unfolded DFG is perfect rate. For example, the least common multiple (lcm) of the number of loop delays in figure (6a) is lcm {2, 3} = 6.

Therefore, unfolding this DFG using unfolding factor 6 guarantees an unfolded DFG with sample period equal to the iteration period. However, we just showed that the unfolding factor 2 can be used to make the sample period equal to the iteration period, so the lcm of the number of loop delays is not necessarily the minimum unfolding factor that can be used to guarantee that the sa2mple period of the unfolded DFG is equal to the iteration period. The lcm is independent of the node computation times.

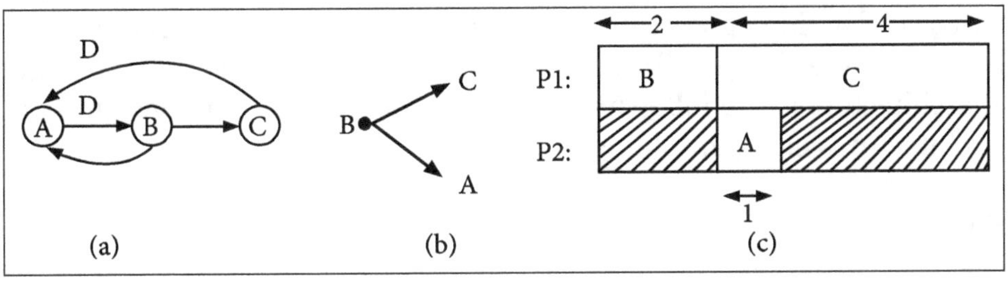

(8) A DFG where execution times of the nodes A, B and C are assumed to be 1, 2 and 4 u.t., respectively, Corresponding acyclic precedence graph and Periodic non-overlapped schedule with period 6 u.t.

Unfolding can be exploited to reduce the iteration period in programmable DSPs. Periodic schedules can be constructed from the acyclic precedence graphs that are obtained by deleting all edges with delay elements from the DFG. If the period of a periodic schedule for a J-unfolded DFG is T, then the average iteration period is T/J.

Schedules can be either,

- Overlapped.
- Non-overlapped.

In a non-overlapped schedule, the period of the schedule is the same as the critical path of the acyclic precedence graph. In an overlapped schedule, the period of the schedule is smaller than the crit2ical path of the acyclic precedence graph.

As an example, consider the DFG in figure (8a) whose acyclic precedence graph and periodic schedule are shown in figure (8b) and figure (8c), respectively. The schedule in this case is non-overlapped. For this DFG, the iteration bound is 3.5 u.t. and the period of the non-overlapped periodic schedule is 6 u.t. Note that the path marked by bold lines in the acyclic precedence graph represents the critical path.

A corresponding 2-unfolded DFG, its acyclic precedence graph and periodic schedule are shown in figure (9). The period of the overlapped periodic schedule is 7 u.t., therefore, the average iteration period is 3.5 u.t., which is the same as the iteration bound. Retiming the DFG in figure (8a) by removing the delay element from the input edge of node B and adding a delay element to each outgoing edge of node B can reduce the critical path from 6 u.t. to 4 u.t. Furthermore, 2-unfolding of this retimed DFG will lead to a non-overlapped schedule with average iteration period, which is the same as the iteration bound. In general, constructing overlapped schedules is significantly more complex than non-overlapped schedules.

Parallel Processing

A direct application of the general unfolding transformation is to design parallel processing architectures from serial processing architectures. At the word level, this means that word-parallel architectures can be designed from word-serial architectures. At the bit level, it means that bit-parallel and digit-serial architectures can be designed from bit-serial architectures. The following techniques describe how to design parallel processing architectures using unfolding:

- Word-level parallel processing.
- Bit-level parallel processing.

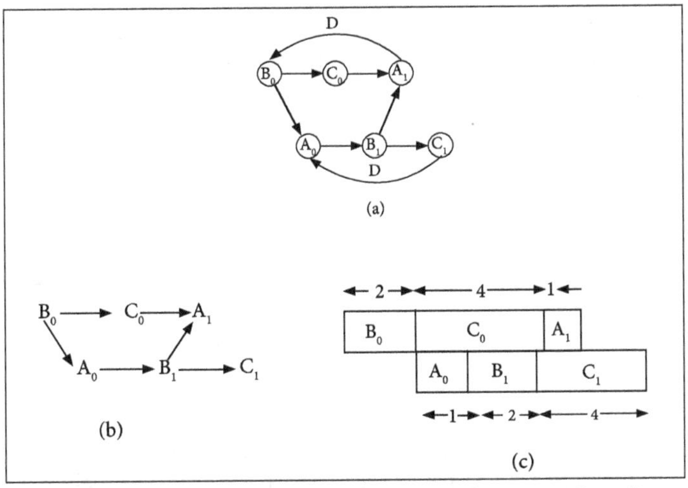

(9) A 2-unfolded DFG, Its acyclic precedence graph and an overlapped periodic schedule with period 7 u.t.

1.7 Folding Transformation

In synthesizing DSP architectures, it is important to minimize the silicon area of the integrated circuits, which is achieved by reducing the number of functional units such as adders and multipliers, registers, multiplexers and interconnection wires. The folding transformation is used to systematically determine the control circuits in DSP architectures, where multiple algorithm operations are time-multiplexed to a single functional unit.

By executing multiple algorithm operations on a single functional unit, the number of functional units in the implementation is reduced, resulting in an integrated circuit with low silicon area. Folding techniques can be used for the synthesis of DSP architectures that can be operated using single or multiple clocks.

Figure (1) shows an example of how 2 addition operations can be time-multiplexed on a single pipelined hardware adder. The DSP in figure (1a) computes y(n) = a(n) + b(n) + c(n). In figure (1b), the 2 addition operations shown in figure (1a) are time-multiplexed on a single pipelined adder. The time-multiplexed hardware in figure (1b) operates as follows:

In cycle 0, the samples a(0) and b(0) are switched into the adder and the sum (a(0) + b(0)) is stored in the delay element until cycle 1, when (a(0) + b(0)) is switched into the adder along with c(0). The sum (a(0) + b(0) + c(0)) is stored in the delay until cycle 2, when the sum (a(0) + b(0)+ c(0)) is output and the intermediate result (a(1) + b(1)) is computed by the adder. This process continues as shown in the below table. One output sample is produced every 2 clock cycles and one sample of each input signal is consumed every 2 clock cycles.

As a result, an input sample must remain valid for two clock cycles before changing. In general, the data on the input of the folded realization is assumed to be valid for N cycles before changing, where N is the number of algorithm operations executed on a single functional unit in hardware.

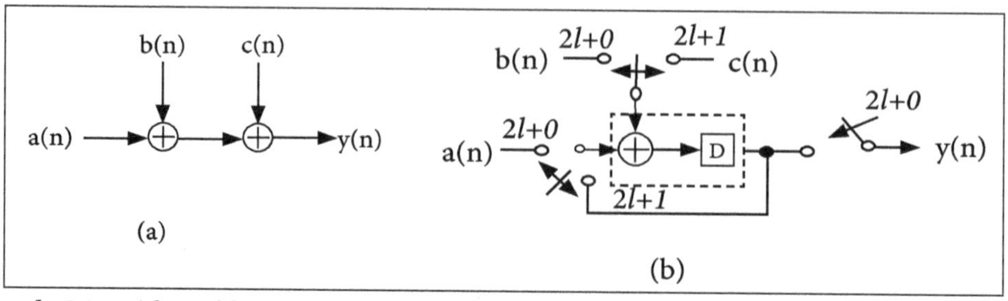

(1) A simple DSP with 2 addition operations and A folded architecture where the 2 addition operations are folded to a single hardware adder with 1 stage of pipelining.

Table: Operation of the first six cycles of the folded hardware in figure (1b):

Cycle	Adder input (Left)	Adder input (Top)	System output
0	a(0)	b(0)	-
1	a(0) + b(0)	c(0)	-
2	a(1)	b(1)	a(0) + b(0) + c(0)
3	a(1) + b(1)	c(1)	-
4	a(2)	b(2)	a(1) + b(1) + c(1)
5	a(2) + b(2)	c(2)	-

The architecture in figure (1b) is simple enough to be designed using ad hoc techniques, however, general DSP equations are more complex than the figure (1a). In such cases, the systematic folding techniques can be used to design the time-multiplexed architectures.

Folding provides a means for trading area for time in DSP architecture. One way to implement the figure (1a) is to use two adders in hardware with a pipelining delay between the two adders. This implementation requires two hardware adders and computes one iteration of the equation in the time required to perform addition, T_{add}. On the other hand, the folded implementation in figure (1b) uses one hardware adder and computes one iteration of the equation in $2T_{add}$ time.

In general, folding can be used to reduce the number of hardware functional units by a factor of N at the expense of increasing the computation time by a factor of N. The two extremes of this are when a fully parallel implementation is used and when a single processor is used.

While the folding transformation reduces the number of functional units in the architecture, it may also lead to an architecture that uses a large number of registers. To avoid architectures using excessive amounts of registers, certain techniques can be used to compute the minimum number of registers required to implement a folded DSP architecture and to allocate data to these registers. Using register minimization along with the folding transformation not only reduces the number of functional units but also keeps the area consumed by memory in the folded architecture to a minimum.

Folding Transformation

The folding transformation provides a systematic technique for designing control circuits for hardware where several algorithm operations are time-multiplexed on a single functional unit.

Consider the edge e connecting the nodes U and V with w(e) delays, as shown in figure (2a). Let the executions of the lth iteration of the nodes U and V be scheduled at the time units Nl + u and Nl + v, respectively, where u and v are the folding orders of the nodes U and V that satisfy $0 \le u, v \le N - 1$. The folding order of a node is the time partition to which the node is scheduled to execute in hardware. The functional units that execute the nodes U and V are denoted as H_U and H_V, respectively.

Note that N is the number of operations folded to a single functional unit and is also referred to as the folding factor. If H_U is pipelined by P_U stages, then the result of the lth iteration of the node U is available at the time unit $Nl+u+P_U$. Since the edge $U \xrightarrow{e} V$ has w(e) delays, the result of the lth iteration of the node U is used by the $(l+w(e))^{th}$ iteration of the node V, which is executed at $N(l+w(e))+v$. Therefore, the result must be stored for,

$$D_F(U \xrightarrow{e} V) = [N(l+w(e)+v)] - [Nl+P_U+u] = N_w(e) - P_U + v - u \quad ...(1)$$

time units, which is independent of the iteration number l.

The edge $U \xrightarrow{e} V$ is implemented as a path from H_U to H_V in the architecture with DF $U \xrightarrow{e} V$ delays and data on this path are input to H_V at Nl + v, as illustrated in figure (2b).

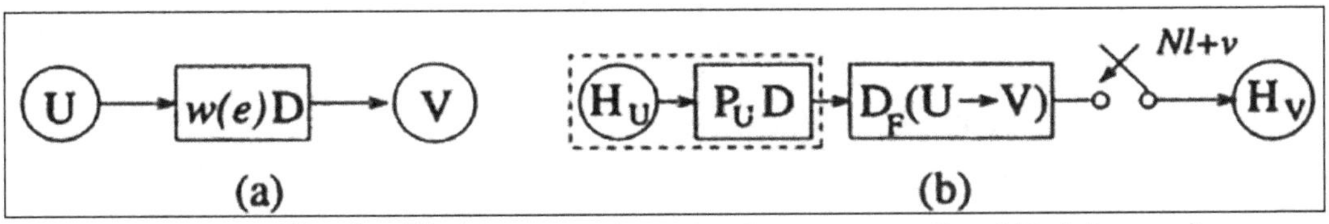

(2) An edge $U \xrightarrow{e} V$ with w(e) delays. (b) The corresponding folded data path.

The data begin at the functional unit H_U which has P_U pipelining stages, pass through $D_F(U \xrightarrow{e} V)$ delays and are switched into the functional unit HV at the time instances $Nl+v$.

A folding set is an ordered set of operations executed by the same functional unit. Each folding set contains N entries, some of which may be null operations. The operation in the jth position within the folding set is executed by the functional unit during the time partition j.

For example, consider the folding set $S_1 = \{A_1, \emptyset, A_2\}$ for N = 3. The operation A_1 belongs to the folding set S_1 with folding order 0 (this is also denoted as $(S_1 | 0)$) and the operation A_2 belongs to the folding set S_1 with folding order 2 (this is also denoted as $(S_1 | 2)$). Due to the null operation in position 1 within S_1, the functional unit that executes operations A_1 and A_2 will not be utilized at time instances $3l + 1$.

The use of systematic folding techniques is demonstrated by folding the retimed biquad filter in figure (3). Assume that addition and multiplication require 1 and 2 u.t., respectively and 1-stage pipelined adders and 2-stage pipelined multipliers are available (i.e., $P_A = 1$ and $P_M = 2$). The functional units can be clocked with period 1 u.t. This filter is folded with folding factor N = 4 using the folding sets shown in the figure.

The folding factor N = 4 means that the iteration period of the folded hardware is 4 u.t., (i.e.) each node of the biquad filter is executed exactly once every 4 u.t. in the folded architecture. It also means that a functional unit in the folded hardware executes four operations of the DSP equation. To see this, the folding sets in figure (3) can be written as $S_1 = \{4, 2, 3, 1\}$ and $S_2 = \{5, 8, 6, 7\}$. The folding set S_1 contains only addition operations and the nodes in this folding set are executed by the same hardware adder.

Similarly, the folding set S_2 contains only multiplication operations and the nodes in this folding set are executed by the same hardware multiplier. For example, node 3 is executed in the folded architecture at time instances 4l + 2 on the adder that implements the operations in the folding set S_1. The folded architecture is shown in figure (4). This is obtained from the Data Flow Graph in figure (3) by writing the folding equation in (1) for each of the 11 edges in the DFG.

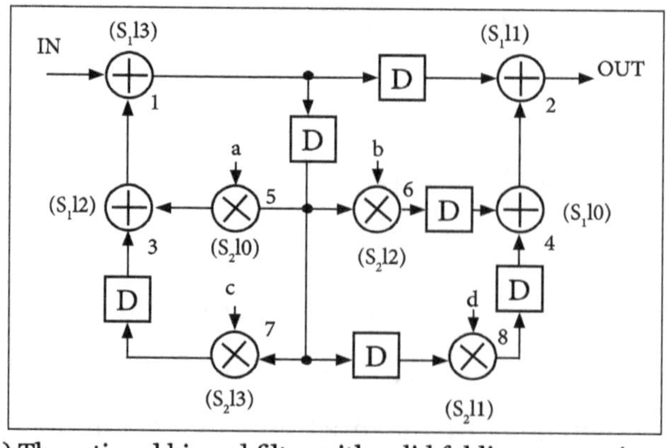

(3) The retimed biquad filter with valid folding sets assigned.

These 11 equations are,

$$D_F(1 \rightarrow 2) = 4(1) - 1 + 1 - 3 = 1$$

$$D_F(1 \rightarrow 5) = 4(1) - 1 + 0 - 3 = 0$$

$$D_F(1 \rightarrow 6) = 4(1) - 1 + 2 - 3 = 2$$

$$D_F(1 \rightarrow 7) = 4(1) - 1 + 3 - 3 = 3$$

$$D_F(1 \rightarrow 8) = 4(2) - 1 + 1 - 3 = 5$$

$$D\ (3 \rightarrow 1) = 4(0) - 1 + 3 - 2 = 0$$

$$D_F(4 \rightarrow 2) = 4(0) - 1 + 1 - 0 = 0$$

$$D_F(5 \rightarrow 3) = 4(0) - 2 + 2 - 0 = 0$$

$$D_F(6 \rightarrow 4) = 4(1) - 2 + 0 - 2 = 0$$

$$D_F(7 \rightarrow 3) = 4(1) - 2 + 2 - 3 = 1$$

$$D_F(8 \rightarrow 4) = 4(1) - 2 + 0 - 1 = 1 \qquad \ldots(2)$$

For example, $D_F(1 \rightarrow 8) = 5$ means that there is an edge from the adder to the multiplier in the folded DFG with 5 delays. Since this edge ends at the node 8, which has folding order 1 in figure (3), the folded edge is switched at the input of the multiplier in the folded DFG at 4l + 1. This folded edge can be seen in figure (4).

For a folded system to be realizable, $D_F\left(U \xrightarrow{e} V\right) \geq 0$ must hold for all of the edges in the DFG. Once valid folding sets have been assigned, retiming can be used to either satisfy this property or determine that the folding sets are not feasible. The edge $U \xrightarrow{e} V$ in figure (2a) is folded using equation (1). Using retiming, the number of delays on the edge $U \xrightarrow{e} V$ is changed from w(e) to,

$$w_r(e) = w(e) + r(V) - r(U) \qquad \ldots(3)$$

Where,

$w_r(e) =$ Number of delays on the edge $U \xrightarrow{e} V$ in the retimed DFG.

$r(X) =$ Retiming value of the node X.

Let, $D'_F\left(U \xrightarrow{e} V\right)$ denote the number of folded delays obtained by folding the edge $U \xrightarrow{e} V$ in the retimed DFG. To ensure that the corresponding edge in the folded hardware has a non-negative number of delays, the constraint $D'_F\left(U \xrightarrow{e} V\right) \geq 0$ must hold, which implies,

$$Nw_r(e) - P_U + v - u \geq 0 \qquad ...(4)$$

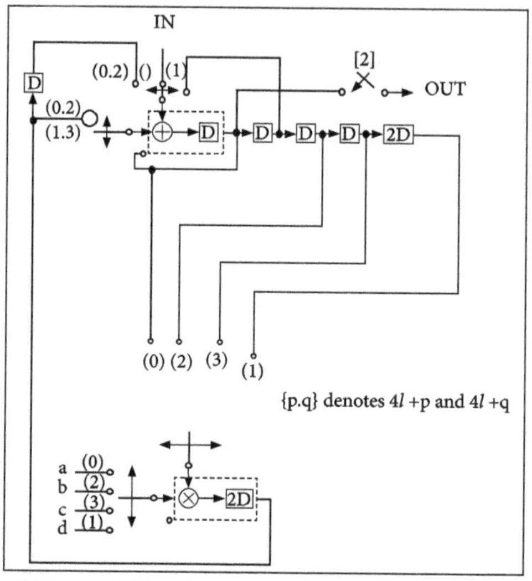

(4) The folded biquad filter using the folding sets given in figure (3).

This constraint ensures that the folding set is valid after retiming. Combining equation (3) and (4) results in,

$$N\left(w(e) + r(V) - r(U)\right) - P_U + v - u \geq 0$$

Substituting $D_F(U \rightarrow V)$ for $N\,w(e) - P_U + v - u$ and solving for r(U) - r(V) results in,

$$r(U) - r(V) \leq \frac{D_F\left(U \xrightarrow{e} V\right)}{N}$$

Since the retiming values of the nodes are restricted to be integers, this can be rewritten as,

$$r(U) - r(V) \leq \left\lfloor \frac{D_F\left(U \xrightarrow{e} V\right)}{N} \right\rfloor \qquad ...(5)$$

Where, $\lfloor x \rfloor$ is the floor of x, which is the largest integer less than or equal to x.

Once the set of constraints for the DFG is found using equation (5), the technique for solving system of inequalities can be used to determine if a solution exists and to find a solution if one indeed exists. As an example, consider the biquad DFG in figure (5).

Assume that 1-stage pipelined adders and 2-stage pipelined multipliers (i.e., $P_A = 1$ and $P_M = 2$) are available. The folding equations and the retiming for folding constraints for the DFG in figure (5a) are given in the below table.

According to the technique for solving systems of inequalities, the 1st step is to draw a constraint graph. The constraint graph for the inequalities in the right column of the below table is shown in figure (6). If this constraint graph contains a negative cycle, then there is no solution to the retiming for folding constraints. If no negative cycle exists, one solution is r(i), which is the shortest path from the node 9 to the node i. A shortest path algorithm, such as the Bellman-Ford algorithm or the Floyd-Warshall algorithm can be used to determine if a negative cycle exists and find a solution if no negative cycle exists.

Using either of these algorithms, we find that the set of constraints has a solution and one solution is r(1) = -1, r(2) = 0, r(3) = -1, r(4) = 0, r(5) = -1, r(6) = -1, r(7) = -2 and r(8) = -1. The retimed DFG is shown in figure (3), the folding equations for the retimed DFG are given in equation (2) and the folded DFG is shown in figure (4).

Table: Folding equations and retiming for folding constraints for the DFG in figure (5a):

Edge	Folding equation	Retiming for folding constraint
1 → 2	$D_F(1 \to 2) = -3$	r(1) - r(2) ≤ -1
1 → 5	$D_F(1 \to 5) = 0$	r(1) - r(5) ≤ 0
1 → 6	$D_F(1 \to 6) = 2$	r(1) - r(6) ≤ 0
1 → 7	$D_F(1 \to 7) = 7$	r(1) - r(7) ≤ 1
1 → 8	$D_F(1 \to 8) = 5$	r(1) - r(8) ≤ 1
3 → 1	$D_F(3 \to 1) = 0$	r(3) - r(1) ≤ 0
4 → 2	$D_F(4 \to 2) = 0$	r(4) - r(2) ≤ 0
5 → 3	$D_F(5 \to 3) = 0$	r(5) - r(3) ≤ 0
6 → 4	$D_F(6 \to 4) = -4$	r(6) - r(4) ≤ -1
7 → 3	$D_F(7 \to 3) = -3$	r(7) - r(3) ≤ -1
8 → 4	$D_F(8 \to 4) = -3$	r(8) - r(4) ≤ -1

The retiming for folding can also be carried out by inspection for simple DFGs such as the biquad filter. From the above table, the edges with negative $D_F(U \rightarrow V)$ are the edges $1 \rightarrow 2$, $6 \rightarrow 4$, $8 \rightarrow 4$ and $7 \rightarrow 3$. Note that adding (subtracting) w delays on the edge $U \rightarrow V$ increases (decreases) the $DF(U \rightarrow V)$ by Nw. Thus, the $D_F(1 \rightarrow 2)$, $D_F(6 \rightarrow 4)$, $D_F(8 \rightarrow 4)$ and $D_F(7 \rightarrow 3)$ can be made non-negative by increasing the number of delays by 1 on each of these edges. This can be achieved by performing cut set retiming for the cut sets marked c_1 and c_2 in figure (5). The retimed DFG obtained in this manner is identical to the DFG in figure (3).

Note that unfolding the folded circuit in figure (4) with unfolding factor 4 leads to the DFG in figure (3), which is the retimed version of the DFG in figure (5). In general, the original DFG and the N-unfolded version of the folded DFG are retimed and pipelined versions of each other. Furthermore, an arbitrary DFG can be unfolded by factor N and the unfolded DFG can be folded with many possible folding sets to generate a family of architectures. An appropriate folding set can be used to obtain the original DFG.

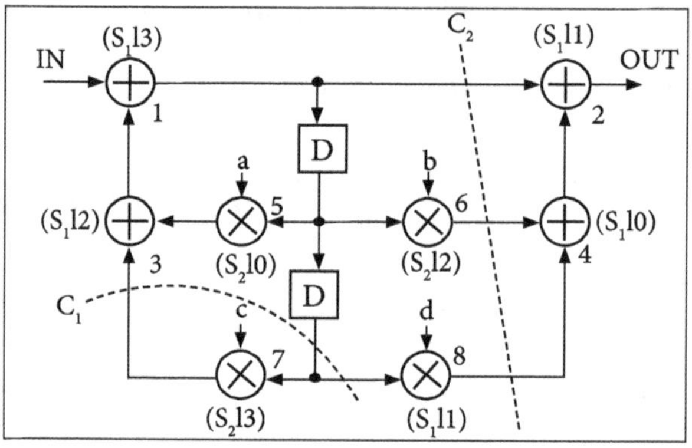

(5) The original DFG resulting in some negative folded edge delays.

The retimed DFG resulting in all non-negative folded edge delays is shown in figure (3).

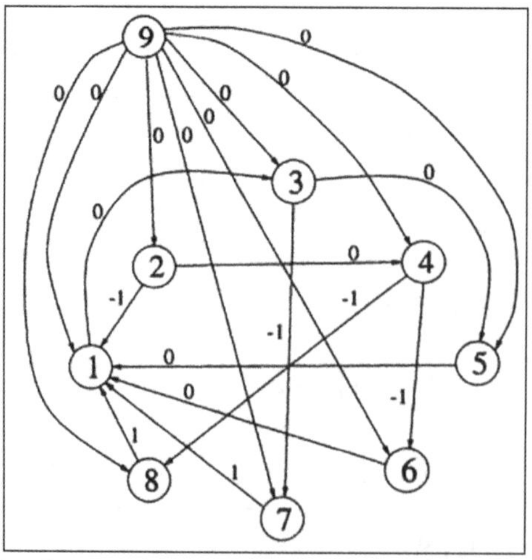

(6) The constraint graph for the set of inequalities in the right-hand column of the above table.

The figure (6), node 9 is the host node. In hardware synthesis, folding can be used for connection binding of the functional units. The folding sets are obtained from a scheduling and allocation algorithm.

1.7.1 Register Minimization Techniques

The focus is to minimize the number of registers used in a DSP architecture, so that the silicon area due to the registers remain small. Techniques are presented for computing the minimum number of registers and for allocating the data to these registers.

Lifetime Analysis

Lifetime analysis is a procedure used to compute the minimum number of registers required to implement a DSP algorithm in hardware. Data sample also called as a variable is live from the time it is produced through the time it is consumed. After the variable is consumed, it is dead. A variable occupies one register during each time unit that it is live. In lifetime analysis, the number of live variables at each time unit is computed and the maximum number of live variables at any time unit is determined. This is the minimum number of registers required to implement the DSP equation.

For example, consider a DSP equation that produces the 3 variables a, b and c. Let the variable a be live during time units $n \in \{1, 2, 3, 4\}$, let the variable b be live during time units $n \in \{2, 3, 4, 5, 6, 7\}$ and let the variable c be live during time units $n \in \{5, 6, 7\}$. Assuming that the lifetimes of variables from the previous and subsequent iterations of the equation do not overlap with the lifetimes of a, b or c, the number of live variables during the time units 1, 2, 3, 4, 5, 6, 7 is 1, 2, 2, 2, 2, 2, 2, respectively. The minimum number of registers required to implement this is the maximum number of live variables at any time unit, which is $\max\{1, 2, 2, 2, 2, 2, 2\} = 2$.

A linear lifetime chart is used to graphically represent the lifetime of each variable in a linear fashion. The linear lifetime chart for our example is shown in figure (7a). The horizontal lines represent clock cycles or time units and the vertical lines represent the lifetimes of the variables. Note that this lifetime chart uses the convention that a variable is not live during the clock cycle in which it is produced and the variable is live during the clock cycle in which it is consumed.

For example, the variable a is produced during cycle 0 and consumed during cycle 4 and this variable is live during the cycles 1, 2, 3, 4. At the right of the lifetime chart, the number of live variables during each cycle is shown. The maximum number of live variables at any time step is 2, so the minimum number of registers that can be used to implement the underlying DSP equation is 2.

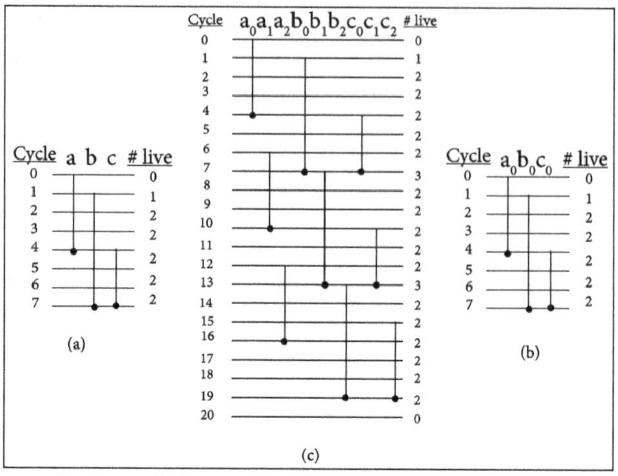

(7) A linear lifetime chart, The linear lifetime chart explicitly showing 3 iterations of the DSP equation assuming the period is N = 6 and The linear lifetime chart implicitly taking into account the periodicity of the DSP equation assuming the period is N = 6.

In general, DSP equations are periodic. The lifetimes of a, b and c from one iteration of the underlying DSP equation may overlap with the lifetimes of these same variables from other iterations. For example, let the iteration period be N = 6 and let a_i, b_i and c_i denote the variables a, b and c resulting from iteration i of the DSP equation. The lifetime chart for three consecutive iterations of the DSP equation is shown in figure (7b). Note that the lifetime of a_1 overlaps with the lifetimes of b_0 and c_0 and as a result the DSP equation requires 3 registers when the periodic nature of the DSP equation is taken into account.

Only two registers are required in figure (7a), where the periodic nature of the DSP equation was not taken into account. This periodic nature of DSP equations must always be taken into account during lifetime analysis. Fortunately, this periodic nature can be taken into account without explicitly drawing the lifetimes of several iterations.

This is done by drawing the lifetimes of the variables for the 0th iteration and letting the number of live variables at the time partitions $n \geq N$ be the sum of the number of live variables due to the 0th iteration at cycles $n - kN$ for all integers $k \geq 0$. This is shown in figure (7c), where the number of live variables in cycle 7 is the sum of the number of live variables due to the 0th iteration at cycles 7 and 1, which is 2 + 1 = 3.

Table: Lifetimes for 3 × 3 matrix transpose operation:

Sample	T_{input}	T_{zlout}	T_{diff}	T_{output}	Life period $(T_{input} \rightarrow T_{output})$
a	0	0	0	4	0 → 4
b	1	3	2	7	1 → 7
c	2	6	4	10	2 → 10
d	3	1	-2	5	3 → 5

e	4	4	0	8	4 → 8
f	5	7	2	11	5 → 11
g	6	2	-4	6	6 → 6
h	7	5	-2	9	7 → 9
i	8	8	0	12	8 → 12

In general, lifetime analysis begins with the construction of a lifetime table, such as the one in the above table. This table represents the transpose operation of a 3 × 3 matrix, (i.e.) the matrix,

$$\begin{bmatrix} a & b & c \\ d & e & f \\ g & h & i \end{bmatrix}$$

is re-ordered from row-by-row to column-by-column ordering.

Each variable in the above table has an input time, T_{input} and a zero-latency output time, T_{zlout}, which is the output time of the variable ignoring the latency of the system. Each variable in the zero-latency system is live for $T_{diff} = T_{zlout} - T_{input}$ clock cycles, however, this violates causality if any of the life periods are negative.

For example, the variable d is input at $T_{input} = 3$ and output at $T_{zlout} = 1$ and is live for 1 - 3 = -2 time units, which is not possible. To force the system to be causal, a latency is added to the system. The latency, T_{lat}, is the magnitude of the most negative value of T_{diff} and $T_{lat} = -(-4) = 4$ in the above table. The actual output time of the variables is $T_{output} = T_{zlout} + T_{lat}$ and the life period of the variables is $\left(T_{input} \rightarrow T_{output}\right)$. Using these life periods, the lifetime chart can be drawn. The lifetime chart for the above table is shown in figure (8). Note that the period of this DSP equation is N = 9.

Another way to represent the periodic behavior of DSP equations using a lifetime chart is to use a circular lifetime chart. In a circular lifetime chart of periodicity N, the point marked i (where i is an integer satisfying 0 < i < N - 1) represents the time partition i and all time instances $\{(Nl + i)\}$, where l is any non-negative integer.

For example, for N = 8, the time partition i = 3 represents the time instances {3, 11, 19, ...}. The circular lifetime chart corresponding to the linear lifetime chart in figure (8) is shown in figure (9). The numbers in parentheses represent the number of live variables at each time partition.

Note that a variable produced during the time unit j and consumed during the time unit k is shown to be alive from the time unit j + 1 to the time unit k because we use the convention that a variable does not need to be stored during the time unit in which it is produced.

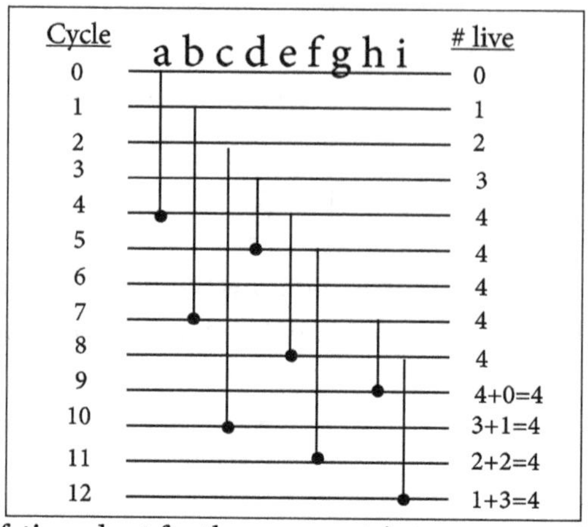

(8) The linear lifetime chart for the 3 × 3 matrix transposer with period N = 9.

Data Allocation using Forward-backward Register Allocation

Once the minimum number of registers required to implement the DSP equation has been determined, the data need to be allocated to these registers. Forward-backward register allocation is an allocation scheme that can be used to allocate data to the minimum number of registers. Forward-backward register allocation is used to generate architectures, that use the minimum number of registers required to implement the DSP equation. Other allocation schemes can also be used.

Register allocation can be performed using an allocation table. The allocation scheme dictates how the variables are assigned to registers in the allocation table. The allocation table that uses the forward-backward scheme to allocate the data for the 3 × 3 matrix transposer is shown in figure (10b).

The steps used to perform forward-backward register allocation are as follows:

Step 1: Determine the minimum number of registers using lifetime analysis.

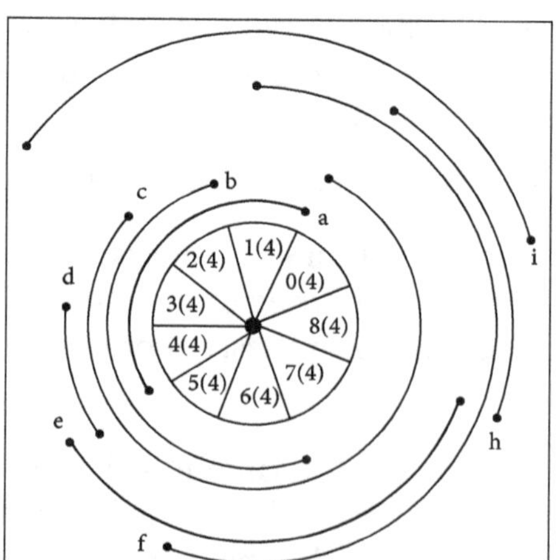

(9) The circular lifetime chart for the 3 × 3 matrix transposer.

The corresponding linear lifetime chart of figure (9) is in figure (8) and the life periods of the variables are given in the above table.

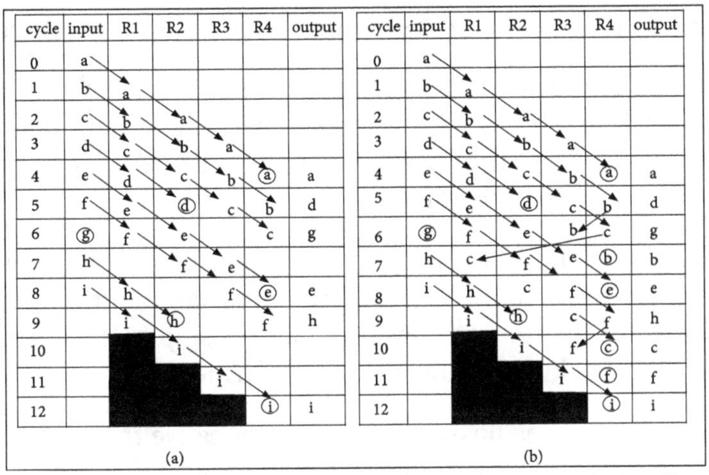

(10) The allocation table for the 3 × 3 matrix transposer and The allocation table after the allocation has been completed.

Step 2: Input each variable at the time step corresponding to the beginning of its lifetime. If multiple variables are input in a given cycle, these are allocated to multiple registers such that the variable with the longest lifetime is allocated to the initial register and the other variables are allocated to consecutive registers in decreasing order of lifetime.

Step 3: Each variable is allocated in a forward manner until it is dead or it reaches the last register. In forward allocation, if the register i holds the variable in the current cycle, then the register i + 1 holds the same variable in the next cycle. If the register i + 1 is not available, then the variable is allocated to the first available forward register.

Step 4: Since the allocation is periodic, the allocation of the current iteration also repeats itself in subsequent iterations. Thus, if R_j is occupied with a variable in cycle l, then R_j would occupy the corresponding variable in cycle l + N, where N denotes the periodicity of the allocation. Therefore, we "hash" the position for R_j at time unit l + N for each j and l.

Step 5: For variables that reach the last register and are not yet dead, the remaining life period is calculated and these variables are allocated to a register in a backward manner on a first-come first-served basis. If multiple registers are available for backward allocation, first try to choose a register such that backward allocation has already been performed from the last register to this register.

In the case, where more than one register qualifies for backward allocation, choose the register with the minimum number of forward registers among all candidate registers that have a sufficient number of forward registers to complete the allocation of the variable. After a variable has been allocated backward, allocate it forward until it is dead or it again reaches the last register.

Step 6: Repeat steps 4 and 5 as required until the allocation is complete.

Figure (10) shows the forward-backward register allocation table for the 3 × 3 matrix transposer whose lifetime table is given in the above table. Step 1 has been performed, where it has been determined that the 3 × 3 matrix transposer requires 4 registers. Steps 2, 3 and 4 are shown in figure (10a).

In step 2, each variable is input at the cycle corresponding to the beginning of its lifetime. Note that, no cycle has more than one input variable in this example. In step 3, the variables are all allocated in a forward manner. In step 4, the hashing is performed to avoid conflicts when backward allocation is performed.

For example, the register R_2 is hashed in cycle 11, because this register is occupied by the variable a in the cycle 11 - 9 = 2. Step 5 is shown in figure (10b). In this step, the variables are allocated backward. For example, the variable b is allocated backward to R_3, because this is the only register available for backward allocation. After each of the variables b, c and f have been allocated backward and then allocated forward until they are dead, the allocation is complete.

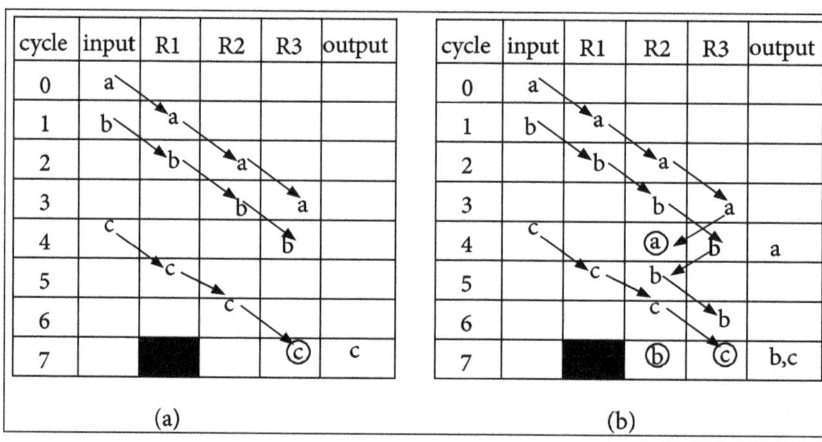

(11) The allocation table for the data represented in figure (7c) and the allocation table after the allocation has been completed.

As another example of forward-backward register allocation, the allocation table for the data represented in figure (7c) is shown in figure (11). Figure (11a) shows the allocation table after steps 1 through 4 of forward-backward register allocation have been performed and figure (11b) shows the table after the allocation has been completed.

From the allocation table, an architecture for the memory unit can be synthesized. The architecture for the 3 × 3 matrix transposer is shown in figure (12). This architecture is completely specified by the allocation table in figure (10b). For example, the backward allocation for the variable b during cycle 5 from R_4 to R_3 in figure (10b) is implemented using a feedback path in figure (12) from the output of R_4 to the input of R_3, which is switched at time $9l+5$.

Similarly, the feedback path for c is implemented using a feedback path from the output of R_4 to the input of R_1, which is switched at time 9l + 6. The switching instance in the architecture for the feedback of the variable f is 9l + 0 because this feedback occurs in figure (10b) during cycle 9 and all switching instances must be of the form 9l + m for $0 \le m \le 8$.

1.7.2 Register Minimization in Folding Architectures

The register minimization techniques can be used to synthesize control circuits in folded architectures that use the minimum possible number of registers. The basic procedure is as follows:

- Perform retiming for folding.
- Write the folding equations.
- Use the folding equations to construct a lifetime table.
- Draw the lifetime chart and determine the required number of registers.
- Perform forward-backward register allocation.
- Draw the folded architecture that uses the minimum number of registers.

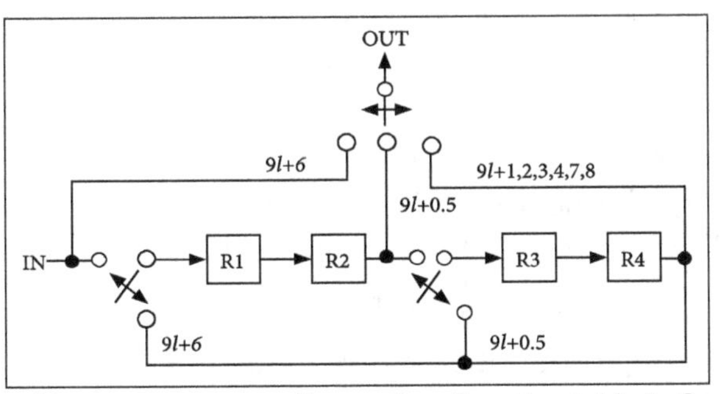

(12) The architecture corresponding to the allocation table in figure (10b).

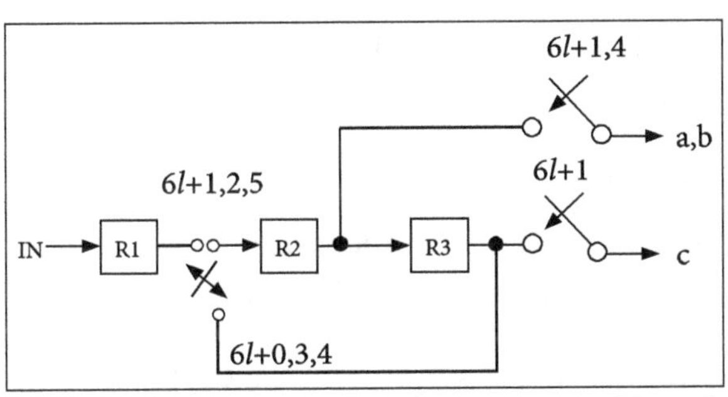

(13) The architecture corresponding to the allocation table in figure (11b).

Example: Biquad Filter:

In this example, minimum-register architecture is constructed for the DFG in figure (5). The DFG after retiming for folding is shown in figure (3) and the folding equations for this DFG are given in equation (2). The folded architecture without any register minimization is shown in figure (4). This architecture uses 6 registers.

Lifetimes for the Retimed Biquad Filter

Node	$T_{input} \rightarrow$ Toutput
1	$4 \rightarrow 9$
2	-
3	$3 \rightarrow 3$
4	$1 \rightarrow 1$
5	$2 \rightarrow 2$
6	$4 \rightarrow 4$
7	$5 \rightarrow 6$
8	$3 \rightarrow 4$

Since retiming for folding has already been performed and the folding equations are written in equation (2), the next step is to construct the lifetime table shown in the above table. In the lifetime table, there is one entry for each node in the DFG that specifies the lifetime $\left(T_{input} \rightarrow T_{output}\right)$ for the node. The time T_{input} for the node U is u + P_U, where u is the folding order of U and P_U is the number of pipelining stages in the functional unit that executes U.

This value of T_{input} is the time unit in which the node produces data in hardware for the 0th iteration of the DSP equation. For example, T_{input} for the node 1 is 3 + 1 = 4. The time T_{output} for the node U is $u + P_U + \max_V \{D_F(U \rightarrow V)\}$, where $\max_V \{D_F(U \rightarrow V)\}$ represents the longest folded path delay among all edges that begin at the node U. This value of T_{output} is the latest time that the result of the 0th iteration of the node is used.

For example, T_{output} for the node 1 is 3 + 1 + max{1, 0, 2, 3, 5} = 3 + 1 + 5 = 9. Note that in this lifetime table it is not necessary to add latency to the output times because the retiming for folding step guarantees the causality of the system.

The linear lifetime chart for the lifetime table is in figure (14). Note that N = 4 is the period of operation. From this lifetime chart, it can be seen that the folded architecture requires two registers. The next step is to perform forward-backward register allocation. The allocation table is shown in figure (15).

The variable n_i, in this table denotes the output of the node i. This table only shows the three variables n_1, n_7 and n_8, because these are the only variables with lifetimes that have non-zero duration. Also, the table shows that the variable n_1 is output in cycle 9. Note that this variable is also output in cycles 4, 5, 6 and 8. The table shows only the latest output time of each variable.

From the allocation table in figure (15) and the folding equations in (2), the final architecture in figure (16) can be synthesized. For example, the edge (1 → 2) has $D_F(1 \to 2) = 1$ delay in the folded architecture. This edge starts at the node 1 and after 1 delay the variable n_1 is located in the register R_1 in figure (15), so there exists an edge from R_1 to the adder in the folded architecture and this edge is switched into the adder at the time instances 4l + 1, because the node 2 has folding order 1.

As another example, the edge (1 → 7) has $D_F(1 \to 7) = 3$ delays and the variable n_1 is in R_2 in after 2 delays, so there is an edge R_2 from to the multiplier in the folded architecture. This edge is switched into the multiplier at the time instances 4l + 3, because node 7 has folding order 3. Note that, in this example, the number of registers in the folded architecture has been reduced from 6 to 2.

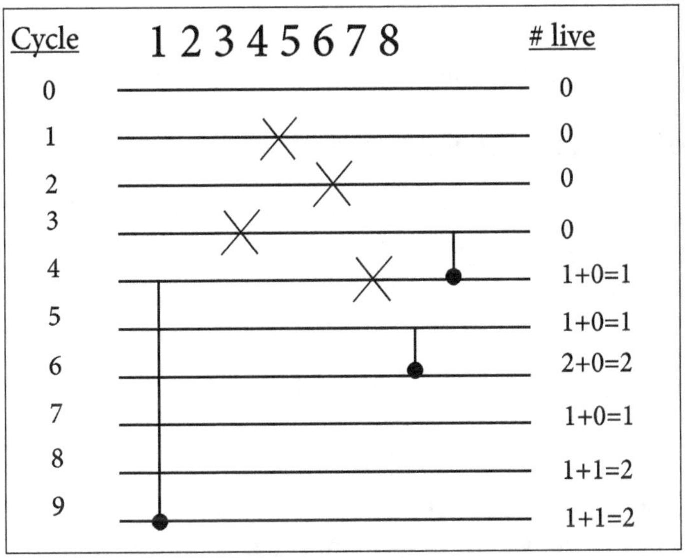

(14) The lifetime chart corresponding to the lifetime table.

cycle	input	R1	R2	output
0				
1				
2				
3	n_8			
4	n_1	n_8		n_8
5	n_7	n_1		
6		n_7	n_1	n_7
7			n_1	
8			n_1	
9			n_1	n_1

(15) The allocation table for the folded biquad filter.

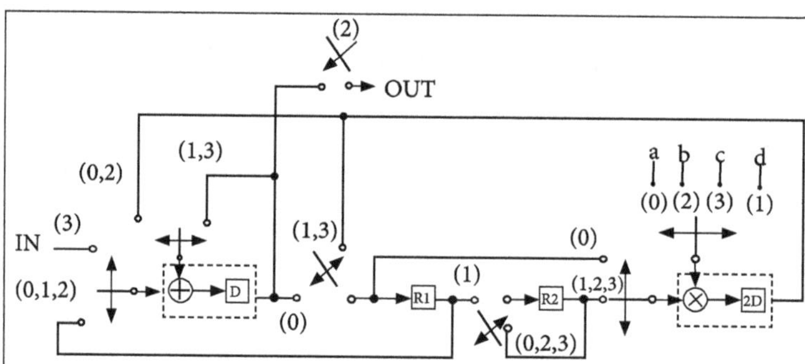

(p.q) denotes 4*l* +p and 4*l*+q

(16) A folded biquad filter architecture implementing the DFG in figure (3) using the minimum number of registers, which is 2.

1.7.3 Folding of Multi Rate Systems

Multi rate DSP algorithm descriptions contain decimators and expanders. The decimator obeys the input-output relationship $y_D(n) = x(Mn)$ and the expander obeys the input-output relationship.

$$yE(n) = \begin{cases} x\left(\frac{n}{M}\right) & \text{if n is a multiple of M} \\ 0 & \text{otherwise} \end{cases}$$

The decimator throws away (M - 1) out of M samples and the expander inserts (M - 1) zero-samples between two non-zero samples. The decimator and the expander both have the effect of changing the sample rate. The folding equation for an edge that contains no decimators or expanders is given in equation (1). Folding equations can also be derived for an edge that contains a decimator or an expander.

Let the lth iteration of the node U be executed at time unit $N_U l + u$ and the lth iteration of V execute at $N_V l + v$, where the folding orders u and v satisfy $u \in [0, N_U)$ and $v \in [0, N_V)$.

The signals are related by,

$$s_1(l) = x(l - w_1)$$

$$s_2(l) = s_1(Ml) = x(Ml - w_1)$$

$$y(l) = s_2(l - w_2) = x(M(l - w_2) - w_1)$$

which implies that the sample y(l), which is consumed during the lth iteration of V, is produced during the $(Ml - (Mw_2 + w_1))^{th}$ iteration of U. The sample y(l) is consumed by H_V in time unit $N_V l + v$ and is computed by H_U in time unit $N_U(Ml - (Mw_2 + w_1)) + u$. If H_U is pipelined by P_U stages, then y(l) is available at time unit $N_U(Ml - (Mw_2 + w_1)) + u + P_U$.

Therefore, y(l) must be stored for,

$$D_F{}^D(U \to V) = [N_V l + v] - [N_U(Ml - (Mw_2 + w_1)) + u + P_U]$$

$$\text{time units} = (N_V - MN_U)l + N_U(Mw_2 + w_1) - P_U + v - u$$

As in the single-rate case, this expression should be independent of the iteration number, l. This can be achieved by forcing $N_V = M\,N_U$, which implies that the node U executes M times for each execution of the node V. The output of the node U is decimated by M before reaching the node V.

With $N_V = M\,N_U$, the folding equation becomes,

$$D_F{}^D(U \to V) = N_U(Mw_2 + w_1)) - P_U + v - u \qquad ...(6)$$

Which is independent of the iteration number, l.

Since the node V is scheduled to execute on hardware operator H_V at time units $N_V l + v$, the data on the edge $U \to V$ are input to H_V at time units $N_V l + v$. For the case of M = 1, (i.e.) where the decimator does not affect the data stream, w_1 and w_2 can be combined as $w = w_1 + w_2$ and $N_U = N_V = N$, where N is the iteration period of the nodes

U and V. Substituting these expressions into equation (6) gives the single-rate folding equation (1), as expected.

As an example of folding edges that contain decimators, consider the DFG in figure (17a). The folding factors or the iteration periods of the nodes are $N_U = 2$ and $N_{V_0} = N_{V_1} = N_{V_2} = N_{V_3} = 6$. The folding orders are $u = 1$, $v_0 = 1$, $v_1 = 2$, $v_2 = 4$ and $v_3 = 5$. Node U is assigned to a processor that is pipelined by one stage, (i.e.) $P_U = 1$.

The folding equations are,

$$D_F^D\left(U \xrightarrow{e_0} V_0\right) = 2\left(3(1)+2\right)-1+1-1=9$$
$$D_F^D\left(U \xrightarrow{e_1} V_1\right) = 2\left(3(0)+1\right)-1+2-1=2$$
$$D_F^D\left(U \xrightarrow{e_2} V_2\right) = 2\left(3(0)+3\right)-1+4-1=8$$
$$D_F^D\left(U \xrightarrow{e_3} V_3\right) = 2\left(3(2)+0\right)-1+5-1=15$$

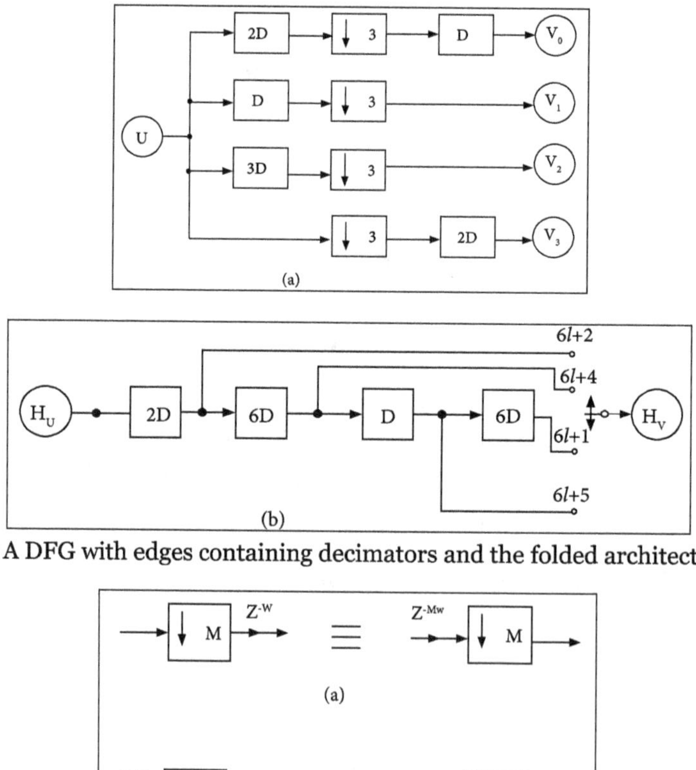

(17) A DFG with edges containing decimators and the folded architecture.

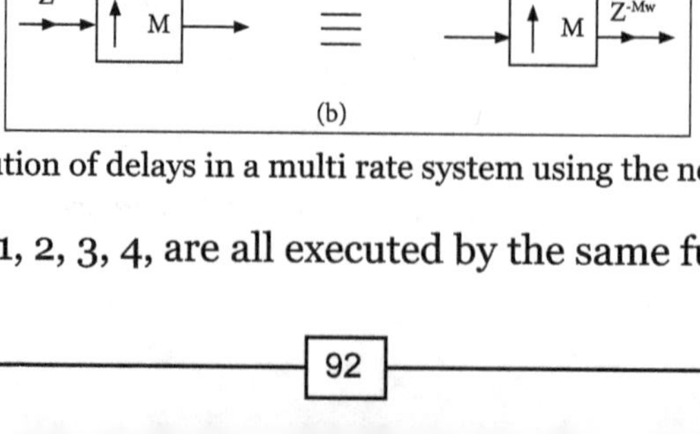

(18) Redistribution of delays in a multi rate system using the noble identities.

If the four nodes V_i, i = 1, 2, 3, 4, are all executed by the same functional unit, H_V, then the folded architecture is shown in figure (17b). The number of registers can be reduced using lifetime analysis.

For equation (6) to be useful, $D_F^{\,D}(U \to V) \geq 0$ must hold a feasible schedule. Retiming for folding can be used for Multi rate DFGs in a manner similar to that used in Single-Rate DFGs.

The noble identities are useful for theory and implementation of multi rate DSP. Special cases of these identities are shown in figure (18). These relationships are used to derive conditions for retiming an MRDFG for folding.

The DFG can be retimed to satisfy the condition $D_F^{\,D}(U \to V) \geq 0$. Let w_1' and w_2' be the number of delays on arc U → V after retiming. Using equation (6), the number of delays on the folded arc after retiming is,

$$D'_F{}^{D}(U \to V) = N_U(Mw_2' + w_1') - P_U + v - u$$

The values of w_1' and w_2' are related to w_1 and w_2 by,

$$w_1' = w_1 + M_r(D_{uv}) - r(U)$$

And,

$$w_2' = w_2 + r(V) - r(D_{uv})$$

Where, r(u) and r(v) are the retiming values of nodes U and V, respectively, (i.e.) the number of times one delay is removed from each of the output arcs of the node and one delay is added to each of the input arcs of the node. According to the noble identity in figure (18a), the retiming value of the decimator, $r(D_{uv})$, is the number of times one delay is removed from its output and M delays are added to its input.

Substituting the expressions for w_1' and w_2', we find,

$$D'_F{}^{D}(U \to V) = N_U\left[M\left(w_2 + r(V) - r(D_{uv})\right) + w_1 + Mr(D_{uv}) - r(U)\right] - P_U + v - u$$

$$= D_F^{\,D}(U \to V) + N_U\left(Mr(V) - r(U)\right)$$

Which is independent of $r(D_{uv})$.

We can retime the DFG for folding by forcing $D'_F{}^{D}(U \to V) \geq 0$, which results in,

$$r(U) - Mr(V) \leq \frac{D_F^D(U \to V)}{N_U}$$

Since M, r(U) and r(V) are all restricted to be integers, this can be written as,

$$r(U)-Mr(V)\leq\frac{D_F^D(U\rightarrow V)}{N_U}$$

Caution must be exercised when retiming a MRDFG due to its periodically time-varying nature. For example, consider the MRDFG in figure (19a). If we retime this DFG by assigning the adder a retiming value of -1 and assigning the multiplier a retiming value of 0, we get the DFG in figure (19b).

The problem is that these two circuits have completely different functionality. In the single-rate case, retiming an input node simply results in a delay at the output signal, where this example shows that retiming an input node of an MRDFG can completely change the functionality of the circuit.

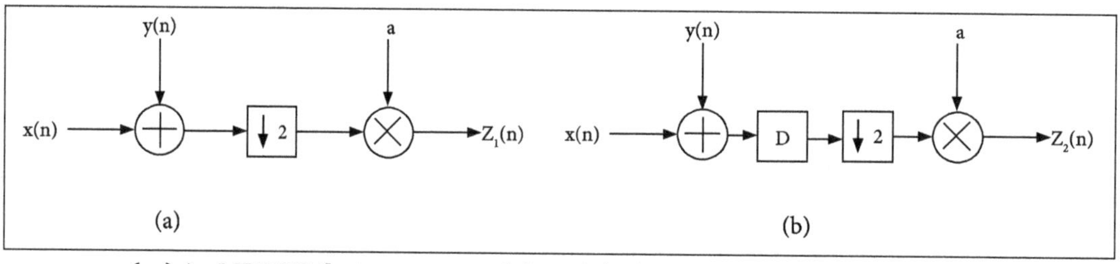

(19) An MRDFG that computes $z_1(n) = a(x(2n) + y(2n))$ and Retimed version that computes $z_2(n) = a(x(2n - 1) + y(2n - 1))$.

2

Wiener Filtering and Spectrum Estimation

2.1 Wiener Filtering

The Wiener filter, which is shown in the below figure, is an important class of optimum linear filter. In its most general form, the Wiener filter consists of an input signal, x(k), a desired response, d(k) and a linear filter with impulse response h(k), which is fed by x(k) to produce the output y(k). The difference between the output of the Wiener filter, y(k) and the desired response, d(k), is denoted by the estimation error, e(k).

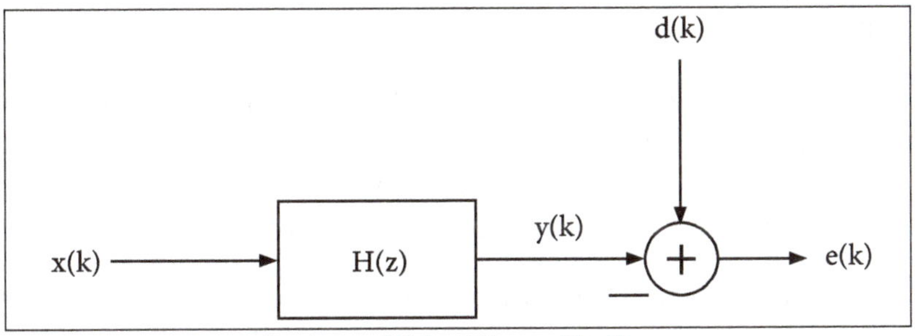

Basic model for Wiener filtering

The objective of Wiener filtering is to determine the impulse response h(k) which makes the estimation error, e(k) small in some statistical sense. However, the one which is mathematically tractable is that, we wish to minimize the mean squared value of e(k). This is what we will use as the index of performance for the Wiener filter.

There are several structures which the Wiener filter is allowed to assume. In each case, there is an optimum solution among filters of each particular class. However, the FIR solution has proved to be the most popular for several reasons, the most significant of which is the ability to implement adaptive solutions to Wiener filtering problem with an FIR structure.

An example of Wiener filtering is shown below:

Consider the communication system shown in the below figure (a). A signal d(k) is transmitted over the communication channel which is described by the transfer function C(z). The model C(z) can incorporate the effects of channel distortion.

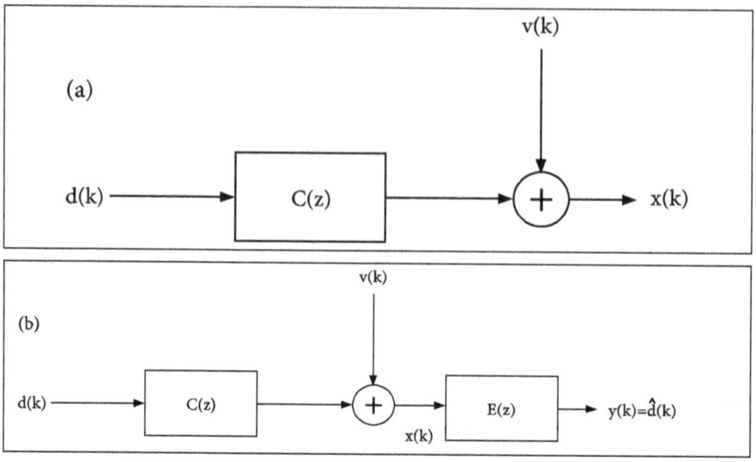

Using a Wiener filter for channel equalization: (a) Model of communication system and (b) Communication system with equalizer.

Further, we assume that, there is channel noise, v(k), which further corrupts the received signal, x(k). We wish to design an equalizer, H(z), which attempts to undo the effects of C(z) and v(k). As shown in the above figure (b), the equalizer has as its input, (i.e.) the received signal x(k) and produces the output y(k). The desired signal is the transmitted signal d(k).

We will consider a Wiener filter, H(z) for the equalizer, which minimizes the mean squared error between y(k) and d(k). Solution of this problem requires a statistical description of the transmitted signal, the channel noise and the received signal. Depending on whether the equalizer is allowed to be causal or non-causal, we will obtain different solutions.

However, for physical implementation we must impose causality. Among the class of causal linear filters, we can allow the equalizer to be FIR or IIR. It may turn out that, the optimal causal linear filter is IIR. In many cases, an FIR filter with enough taps may provide an adequate solution.

2.1.1 The FIR Wiener Filter-Filtering, Linear Prediction, Noise Cancellation

FIR Wiener filter

Suppose x(n) is a WSS process that is correlated with the WSS process d(n) and we want to estimate d(n) from x(n).

Suppose the estimate $\hat{d}(n)$ is obtained by applying an FIR filter w(n) on x(n), then,

$$\hat{d}(n)=\sum_{l=0}^{p-1} w(l)x(n-l)=\left[w(0),\ldots,w(p-1)\right]\begin{bmatrix} x(n) \\ \vdots \\ x(n-p+1) \end{bmatrix}=w^{T}x$$

The filter co-efficient can now be found by minimizing the mean square error.

$$\xi = E\left\{|e(n)|^2\right\} = E\left\{\left|\hat{d}(n) - d(n)\right|^2\right\} = E\left\{\left|w^T x - d(n)\right|^2\right\}$$

$$= w^T \underbrace{E\{xx^H\}}_{R_x^*} w^* - \underbrace{E\{d(n)x^H\}}_{r_{dx}^T} w^* - w^T \underbrace{E\{xd^*(n)\}}_{r_{dx}^*} + \underbrace{E\left\{|d(n)|^2\right\}}_{r_d(0)}$$

$$= w^H R_x w - w^H r_{dx} - r_{dx}^H w + r_d(0)$$

Note that, R_x is the auto-correlation matrix containing $r_x(k) = E\{x(n)x^*(n-k)\}$ as entries and r_{dx} is the cross-correlation vector containing $r_{dx}(k) = E\{d(n)x^*(n-k)\}$ as entries.

We can find the optimal filter coefficients by setting the gradient towards w* to zero.

$$\Delta_{w^*}\xi(w, w^*) = R_x w - r_{dx} = 0 \Rightarrow R_x w = r_{dx}$$

These are the so-called Wiener-Hopf equations which can be solved easily.

It can be shown that the resulting error is uncorrelated with the known data.

$$E\{e(n)x^H\} = E\left\{\left(w^T x - d(n)\right)x^H\right\} = w_T\ R_x{}^* - r_{dx}{}^T = 0$$

This is known as the orthogonality principle or projection theorem.

The resulting error is given by,

$$\xi_{min} = r_d(0) - r_{dx}{}^H w = r_d(0) - r_{dx}{}^H R_x{}^{-1} r_{dx}$$

Filtering

We consider the following data model:

$$x(n) = d(n) + v(n) \text{ or } x = d + v$$

Where,

$$d = [d(n), \ldots, d(n-p+1)]^T$$

$$v = [v(n), \ldots, v(n-p+1)]^T$$

This results in the following expressions for R_x and r_{dx}:

$$R_x = E\{x^* x^T\} = E\left\{(d+v)^*(d+v)^T\right\} = E\{d^* d^T\} + E\{v^* v^T\} = R_d + R_v$$

$$r_{dx} = E\{d(n)x^*\} = E\{d(n)d^*\} = r_d$$

Then, the Wiener-Hopf equation becomes as given below,

$$(R_d + R_v)w = r_d$$

Prediction

We consider the following data model:

$$d(n) = x(n+\alpha)$$

This results in the following expression for r_{dx}:

$$r_{dx} = E\{d(n)x^*\} = E\{x(n+\alpha)x^*\} = r\alpha$$

Then, the Wiener-Hopf equation becomes,

$$\underbrace{\begin{bmatrix} r_x(0) & r_x^*(1) & \cdots & r_x^*(p-1) \\ r_x(1) & r_x(0) & \cdots & r_x^*(p-2) \\ \vdots & \vdots & \ddots & \vdots \\ r_x(p-1) & r_x(p-2) & \cdots & r_x(0) \end{bmatrix}}_{R_x} \underbrace{\begin{bmatrix} w(0) \\ w(1) \\ \vdots \\ w(p-1) \end{bmatrix}}_{w} = \underbrace{\begin{bmatrix} r_x(\alpha) \\ r_x(\alpha+1) \\ \vdots \\ r_x(\alpha+p-1) \end{bmatrix}}_{r_\alpha}$$

For $\alpha = 1$, this corresponds to the all-pole modeling using Prony's method, the auto-correlation method or the Yule-Walker method.

Prediction in Noise

We consider the following data model:

$$y(n) = x(n) + v(n) \text{ or } y = x + v \text{ and } d(n) = x(n+\alpha)$$

Where,

$$y = [y(n), \ldots, y(n-p+1)]^T$$

$$v = [v(n), \ldots, v(n-p+1)]^T$$

This results in the following expressions for R_y and r_{dy}:

$$R_y = E\{y^* y^T\} = E\{(x+v)^*(x+v)^T\} = E\{x^* x^T\} + E\{v^* v^T\} = R_x + R_v$$

$$r_{dy} = E\{d(n)y^*\} = E\{x(n+\alpha)(x^* + v^*)\} = E\{x(n+\alpha)x^*) = r_\alpha$$

Then, the Wiener-Hopf equation becomes,

$$(R_x + R_v)w = r_\alpha$$

Noise Cancellation

We consider the same data model as for filtering.

$$x(n) = d(n) + v_1(n) \text{ or } x = d + v_1$$

Where,

$$d = [d(n), \ldots, d(n-p+1)]^T$$

$$v_1 = [v_1(n), \ldots, v_1(n-p+1)]^T$$

This time we estimate $v_1(n)$ from a correlated noise source $v_2(n)$ and estimate d(n) as,

$$\hat{d}(n) = x(n) - \hat{v}_1(n) \text{ with } \hat{v}_1(n) = w^T v_2$$

Where, $v_2 = [v_2(n), \ldots, v_2(n-p+1)]^T$

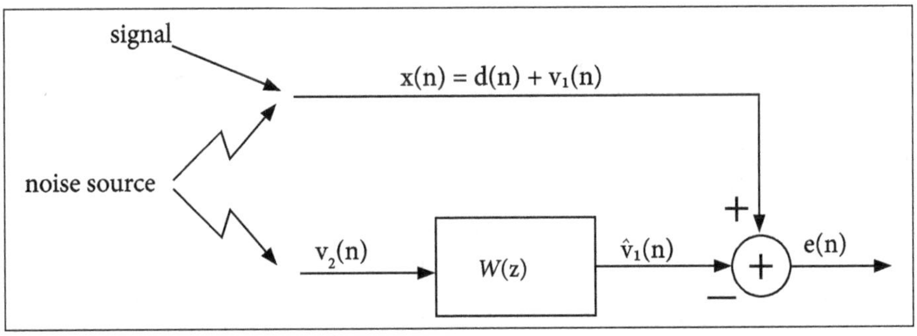

To estimate $v_1(n)$ from $v_2(n)$, we start from the Wiener-Hopf equations,

$$R_{v2}w = r_{v1v2}$$

Since, r_{v1v2} is generally not known, we can rewrite this as,

$$r_{v1v2} = E\{v_1(n)v_2^*\} = E\{(d(n) + v_1(n))v_2^*\} = E\{x(n)v_2^*\} = r_{xv2}$$

Thus, the Wiener-Hopf equations can be written as,

$$R_{v2}w = r_{xv2}$$

d(n) is then estimated as,

$$\hat{d}(n) = x(n) - \hat{v}_1(n) \text{ with } \hat{v}_1(n) = w^T v_2$$

2.2 IIR Wiener Filter

If we do not put a constraint on the length of the filter, it is generally infinitely long and non-causal.

The Wiener-Hopf equations are the same as before, but with an infinitely long sum.

$$\sum_{l=-\infty}^{\infty} r_x(k-l)h(l) = r_{dx}(k); \quad -\infty < k < \infty$$

In contrast to the FIR case, the LHS is the convolution between rx(k) and h(k).

$$r_x(k) * h(k) = r_{dx}(k) \Leftrightarrow P_x(z)H(z) = P_{dx}(z)$$

So, the system function of the IIR Wiener filter is given by,

$$H(z) = \frac{P_{dx}(z)}{P_x(z)}$$

Using the spectral factorization of $P_x(z) = \sigma_0{}^2 Q(z)Q^*(1/z^*)$, this can also be written as,

$$H(z) = \frac{P_{dx}(z)}{\sigma_0^2 Q(z)Q^*(1/z^*)}$$

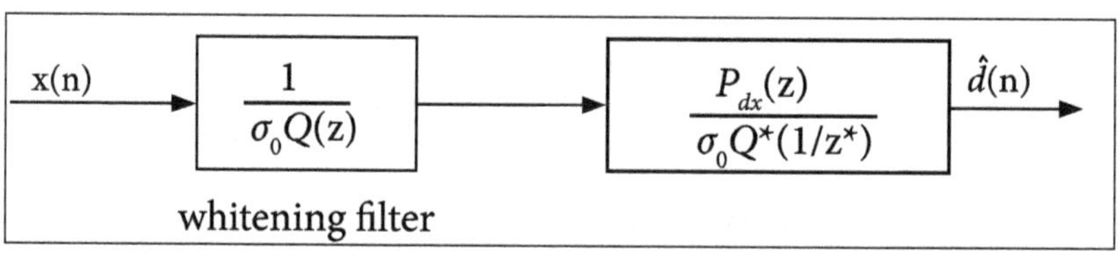

The minimum error is the same as before, but again with an infinitely long sum.

$$\xi_{min} = r_d(0) - \sum_{l=-\infty}^{\infty} h(l) r_{dx}^*(l)$$

Using Parseval's theorem, this error can also be written as,

$$\xi_{min} = \frac{1}{2\pi}\int_{-\pi}^{\pi}\left[P_d(e^{jw}) - H(e^{jw})P_{dx}^*(e^{jw})\right]d\omega$$

Non-causal IIR Wiener Filter

The non-causal IIR filter has a system function which is given by,

$$H(z)=\sum_{k=-\infty}^{\infty} h(k)z^{-k} \qquad \text{...(1)}$$

This contains the components with the negative powers of z as well as with the positive powers of z. This means that H(z) contains both causal part, $\left[H(z)\right]_{+}=\sum_{k=0}^{\infty} h(k)z^{-k}$ and non-causal part $\left[H(z)\right]_{-}=\sum_{k=-\infty}^{-1} h(k)z^{-k}$. A non-causal IIR filter is a smoothing filter and is unrealizable.

Let us consider a non-causal IIR Wiener filter, that we use to estimate the desired signal d(n) from a noisy signal x(n). It is assumed that d(n) and x(n) are jointly WSS with given auto-correlations $r_d(k)$ and $r_x(k)$ and the cross-correlation $r_{dx}(k)$.

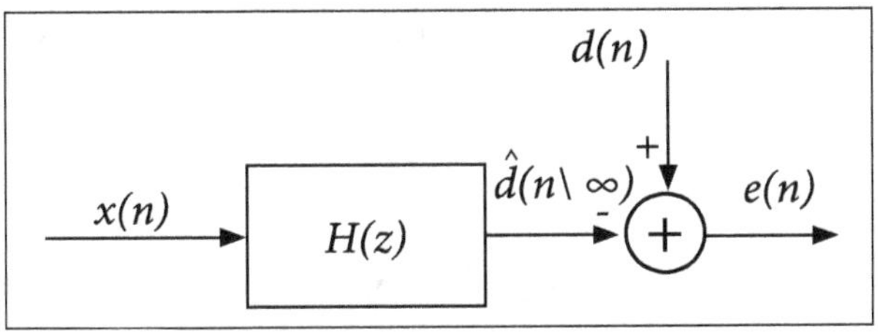

A non-causal IIR Wiener filter.

The input to the filter is the noisy signal x(n) and the output is,

$$\hat{d}(n\,|\,\infty)=\sum_{l=-\infty}^{\infty} h(l)x(n-l) \qquad \text{...(2)}$$

which is the estimate of d(n) at time n based on all the signal values x(n) up to the infinite time instant $n=\infty$, (i.e.) x(n) for $-\infty<n<\infty$.

The estimate error is given by,

$$e(n)=d(n)-\hat{d}(n\,|\,\infty)=d(n)-\sum_{l=-\infty}^{\infty} h(l)x(n-1) \qquad \text{...(3)}$$

To design an optimum IIR Wiener filter with system function H(z) in equation (1) or equivalently to determine the filter co-efficient h(l) for $-\infty<l<\infty$ that produce the minimum mean square error, we use the three step optimization.

We first set $\partial\xi/\partial h^*(k)=0$ for $-\infty<k<\infty$ and then we have,

$$\frac{\partial\xi}{\partial h^*(k)}=\frac{\partial}{\partial h^*(k)}E\left\{\left|e(n)\right|^2\right\}=\frac{\partial}{\partial h^*(k)}E\left\{e(n)e^*(n)\right\}$$

$$= E\left\{e(n)\frac{\partial}{\partial h^*(k)}\left[d^*(n) - \sum_{k=-\infty}^{\infty} h^*(k)x^*(n-k)\right]\right\}$$

$$= -E\{e(n)x^*(n-k)\} = 0, \quad -\infty < k < \infty \qquad ...(4)$$

Which is the orthogonality principle.

Substituting e(n) into equation (4), we have,

$$E\left\{\left[d(n) - \sum_{l=-\infty}^{\infty} h(l)x(n-l)\right]x^*(n-k)\right\}$$

$$= E\{d(n)x^*(n-k)\} - \sum_{l=-\infty}^{\infty} h(l)E\{x(n-l)x^*(n-k)\} = 0 \qquad ...(5)$$

Using the relations $r_{dx}(k) = E\{d(n)x^*(n-k)\}$ and $r_x(k-l) = E\{x(n-l)\,x^*(n-k)\}$ in equation (5), we obtain the Wiener-Hopf equations for the non-causal IIR Wiener filter.

$$\sum_{l=-\infty}^{\infty} h(l)r_x(k-l) = r_{dx}(k); \quad -\infty < k < \infty \qquad ...(6)$$

Obviously, for an infinite number of coefficients, we cannot use the matrix operation. However, on the left side of equation (6), there is a convolution of h(n) with $r_x(k)$, so that we have,

$$h(k) * r_x(k) = r_{dx}(k) \qquad ...(7)$$

Which becomes $H(z)P_x(z) = P_{dx}(z)$ in the z-transform domain.

Thus, the system function of the IIR filter may be written as the ratio of the cross-power spectrum $P_{dx}(z)$ and the power spectrum of $x(n)$, $P_x(z)$,

$$H(z) = \frac{P_{dx}(z)}{P_x(z)} \qquad ...(8)$$

This shows that the coefficients h(l) of an IIR Wiener filter can be found by the inverse z-transform of $H(z) = P_{dx}(z)/P_x(z)$, where, $P_{dx}(z)$ and $P_x(z)$ are available from the z-transform of $r_x(k)$ and $r_{dx}(k)$.

In the same way, the minimum mean square error can be obtained as follows:

$$\xi_{min} = E\{e(n)d^*(n)\} = r_d(0) - \sum_{l=-\infty}^{\infty} h(l)r_{dx}^*(l) \qquad ...(9)$$

Using Parseval's theorem and considering $r_d(0)=(1/2\pi)\int_{-\pi}^{\pi}P_d(e^{j\omega})d\omega$ the error, this can be expressed as,

$$\xi_{min}=r_d(0)-\frac{1}{2\pi}\int_{-\pi}^{\pi}H(e^{j\omega})P_{dx}^{*}(e^{j\omega})d\omega=\frac{1}{2\pi}\int_{-\pi}^{\pi}\left[P_d(e^{j\omega})-H(e^{j\omega})P_{dx}^{*}(e^{j\omega})\right]d\omega \quad ...(10)$$

If a noisy signal is of the form $x(n)=h_s(n)*d(n)+v(n)$ in which d(n) and v(n) are uncorrelated with the given $r_d(k)$ and $r_v(k)$, then in equations (8) and (9) we can use equations $P_x(z)=H_s(z)\ H_s{}^*(1/z^*)P_d(z)+P_v(z)$ and $P_{dx}(z)=H_s{}^*(1/z^*)\ P_d(z)$, to determine H(z).

If we have x(n) = d(n) + v(n), then we use $P_x(z)=P_d(z)+P_v(z)$ and $P_{dx}(z)=P_d(z)$, instead. In this case, for example, we have the Wiener-Hopf equations,

$$h(k)*\left[r_d(k)+r_v(k)\right]=r_d(k) \quad ...(11)$$

The z-transform gives us the system function,

$$H(z)=\frac{P_d(z)}{P_d(z)+P_v(z)} \quad ...(12)$$

The minimum mean square error is given by,

$$\xi_{min}=\frac{1}{2\pi}\int_{-\pi}^{\pi}\left[(e^{j\omega})-H(e^{j\omega})P_{dx}^{*}(e^{j\omega})\right]d\omega=\frac{1}{2\pi}\int_{-\pi}^{\pi}P_d(e^{j\omega})\left[1-H(e^{j\omega})\right]d\omega$$

$$=\frac{1}{2\pi}\int_{-\pi}^{\pi}P_d(e^{j\omega})\left[\frac{P_v(e^{j\omega})}{P_d(e^{j\omega})+P_v(e^{j\omega})}\right]d\omega=\frac{1}{2\pi}\int_{-\pi}^{\pi}P_v(e^{j\omega})H(e^{j\omega})d\omega \quad ...(13)$$

2.2.1 The Causal IIR Wiener Filter

Since the IIR Wiener filter is generally non-realizable, we can put a causal constraint. The Wiener Hopf equations are the same, but with a one sided infinitely long sum.

$$\sum_{l=0}^{\infty}r_x(k-1)h(l)=r_{dx}(k);\ 0\le k<\infty$$

However, now the LHS is NOT equal to the convolution between $r_x(k)$ and h(k). To solve this problem, let us assume that x(n) is a unit variance white noise є(n) and thus $r_x(k)=r_\varepsilon(k)=\delta(k)$ and $r_{dx}(k)=r_{d\varepsilon}(k)$. Denoting the filter as g(n), we then obtain,

$$g(n)=r_{d\varepsilon}(n)u(n)\Leftrightarrow G(z)=\left[P_{d\varepsilon}(z)\right]_+$$

So, we can find the causal IIR Wiener filter for x(n) by first whitening x(n) with a causal whitening filter and then applying the above causal IIR Wiener filter for ϵ(n).

To whiten x(n), we first perform the spectral factorization of $P_x(z)$ as,

$$P_x(z) = \sigma_o^2 Q(z) Q^*(1/z^*)$$

Where,

$Q(z) = 1 + q(1)z^{-1} + q(2)z^{-2} + \ldots = N(z)/D(z)$, which is stable and has a minimum phase.

Then, the whitening filter with input x(n) and output ϵ(n) is given by,

$$F(z) = \frac{1}{\sigma_o Q(z)}$$

Since, we can write the power spectrum of $^o(n)$ as,

$$P_\varepsilon(z) = P_x(z)\, F(z)\, F^*(1/z^*) = 1$$

So, the overall causal IIR Wiener filter is given by,

$$H(z) = F(z)G(z) = \frac{\left[P_{d\epsilon}(z)\right]_+}{\sigma_o Q(z)}$$

The problem of finding $P_{d\varepsilon}(z)$ depends on F(z), since $^o(n)$ is formed by filtering x(n) by f(n). Let us rewrite $r_{d\varepsilon}(k)$.

$$r_{d\varepsilon}(k) = E\{d(n)\epsilon^*(n-k)\} = E\left\{d(n)\sum_{l=-\infty}^{\infty} f^*(l)x^*(n-k-1)\right\}$$

$$= \sum_{l=-\infty}^{\infty} f^*(l) r_{dx}(k+l) = \sum_{l=-\infty}^{\infty} f^*(-l) r_{dx}(k-l)$$

As a result, we can write $P_{d\varepsilon}(z)$ as,

$$P_{d\epsilon}(z) = P_{dx}(z)F^*(1/z^*) = \frac{P_{dx}(z)}{\sigma_o Q^*(1/z^*)}$$

Using this expression, we obtain,

$$H(z) = \frac{1}{\sigma_o^2 Q(z)}\left[\frac{P_{dx}(z)}{Q^*(1/z^*)}\right]_+$$

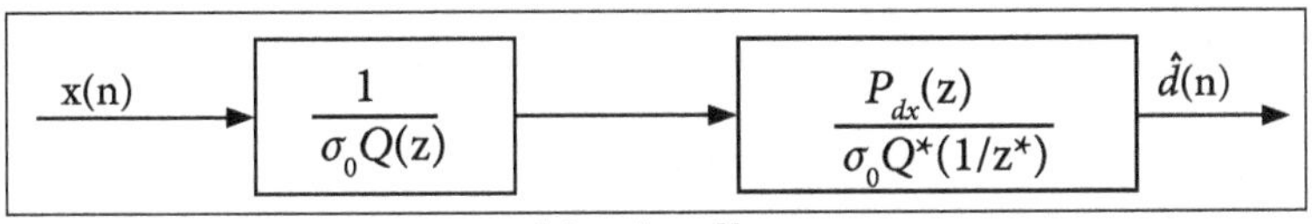

whitening filter.

The minimum error is the same as before, but again with a one sided infinitely long sum.

$$\xi_{min} = r_d(0) - \sum_{l=-\infty}^{\infty} h(l) r_{dx}^*(l)$$

Using Parseval's theorem, this error can also be written as,

$$\xi_{min} = \frac{1}{2\pi}\int_{-\pi}^{\pi}\left[P_d\left(e^{j\omega}\right) - H\left(e^{j\omega}\right)P_{dx}^*\left(e^{j\omega}\right)\right]d\omega$$

2.2.2 Causal Wiener Filtering

Assume that x and y are two jointly stationary second order processes, with spectral density given by,

$$\Phi(z) = \begin{bmatrix} \Phi_x & \Phi_{xy} \\ \Phi_{yx} & \Phi_y \end{bmatrix}(z)$$

We would like to find a strictly causal function F(t), such that $\chi(t) = F(z)y(t)$ is the best linear estimator of x(t) given, y(t - 1), y(t - 2), ...

If y(t) is a white noise process, it is easy to see that the optimal filter is given by a causal function,

$$\hat{F}(z) = \left[\Phi_{xy}(z)\right]_+$$

Where, $[\cdot]_+$ denotes the strictly causal part of the function.

Lemma: Assume the observation process y is a normalized white noise. Then the matrix function F defining the best linear causal estimator $\chi(t)$ of x(t) given the past history of y up to and including time t - 1, (i.e.),

$$\chi(t) := E\left[x(t)H_t^-(y)\right] = \sum_{s=-\infty}^{\infty} F(t-s)y(s)$$

is given by,

$$F(t) = \begin{Bmatrix} \Lambda_{xy}(t), & t > 0 \\ 0, & t \le 0 \end{Bmatrix}$$

Where, Λ_{xy} is the cross covariance matrix of the processes x and y.

Proof: The orthogonality condition provides,

$$E\left\{\left(x(t)-\chi(t)\right)y(\tau)'\right\}=E\left\{\left(x(t)-\sum_{s=\infty}^{\infty}F(t-s)y(s)\right)y(\tau)'\right\}=0,\quad \tau\le t-1$$

Which can be written as,

$$\Lambda_{xy}(t-\tau)=\sum_{s-=-\infty}^{\infty}F(t-s)I\delta(t-s)=F(t-\tau),\quad \tau\le t-1$$

And in order that $\chi(t)\in H_t^-(y)$, the function F has to be strictly causal.

In general, the process y is not a white noise and then the following method using a whitening filter can be applied. Assuming that we can determine a minimum phase spectral factor W of ϕ_y, the process $e(t)=W^{-1}(z)y(t)$ is the normalized innovation process of y(t), (i.e.) a white noise process and we can consider the cascade form of the Wiener filter illustrated in the below figure.

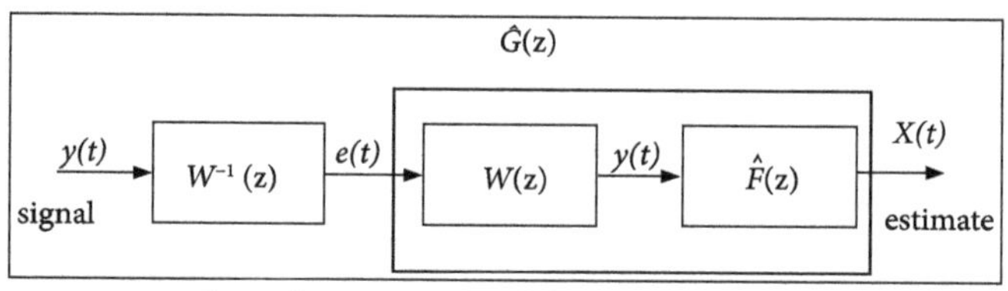

Cascade structure of the Wiener filter.

The part $\hat{G}$ can now be determined as above, the only difference is that we have to exchange ϕ_{xy} with,

$$\phi_{xe}=\phi_{xy}\,W_-^{*} \qquad \ldots(1)$$

The estimator is then given by,

$$\chi(t)=\frac{1}{W(z)}\left[\frac{\Phi_{xy}(z)}{W^{*}(z)}\right]_+ y(t)$$

To derive the expression (1), assume that $\epsilon(t)$ is the innovation process of the joint process (x(t), y(t)). Then consider the spectral representations,

$$x(t)=\int_{-\pi}^{\pi}e^{it\theta}d\hat{x}(\theta)=\int_{-\pi}^{\pi}e^{it\theta}W_x\left(e^{i\theta}\right)d\hat{\epsilon}(\theta)$$

And,

$$y(t)=\int_{-\pi}^{\pi}e^{it\theta}d\hat{y}(\theta)=\int_{-\pi}^{\pi}e^{it\theta}W_y\left(e^{i\theta}\right)d\hat{\epsilon}(\theta)$$

The cross-covariance, $\Lambda_{xy}(\tau) = E\{x(t+\tau)y(t)\}$, can be written as,

$$\Lambda_{xy}(\tau) = E\left\{\int_{-\pi}^{\pi} e^{i(t+r)\theta} W_x(e^{i\theta}) d\in(\theta) \int_{-\pi}^{\pi} e^{-it\theta} W_y(e^{-i\theta}) d\in(\theta)\right\}$$

$$= \int_{-\pi}^{\pi} e^{i\tau\theta} W_x(e^{i\theta}) W_y(e^{-i\theta}) \frac{d\theta}{2\pi},$$

Thus, $\phi_{xy} = W_x W_y{}^*$. Now, since $e(t) = W^{-1}(z)y(t) = W^{-1}(z)W_y(z)\,\varepsilon(t)$ it holds that,

$$\phi_{xe} = W_x\left(W^{-1}\,W_y\right)^* = W_x W_y{}^* W^{-*} = \phi_{xy} W^{-*}$$

Which proves equation (1).

2.2.3 Causal Linear Prediction

Since the infinite past is now being used to predict the next value of x(n), we expect a smaller mean square prediction error than for an FIR linear predictor.

$$\hat{x}(n+1) = \sum_{k=0}^{\infty} h(k)x(n-k)$$

For linear prediction, d(n) = x(n + 1) and the cross-correlation between x(n) and d (n) is given by,

$$r_{dx}(k) = E\{d(n)\,x^*(n-k)\} = E\{x(n+1)x^*(n-k)\} = r_x(k+1)$$

Therefore, $P_{dx}(z) = zP_x(Z)$ and the causal Wiener predictor is given by,

$$H(z) = \frac{1}{\sigma_0^2 Q(z)}\left[\frac{zP_x(z)}{Q^*(1/z^*)}\right]_+ \qquad ...(1)$$

However, since $P_x(z) = \sigma_0^2\,Q(z)Q^*(1/z^*)$, then equation (1) may be simplified as follows:

$$H(z) = \frac{1}{Q(z)}\left[z\,Q(z)\right]_+ \qquad ...(2)$$

Q(z) is a monic polynomial.

$$Q(z) = 1 + q\,(1)z^{-1} + q(2)z^{-2} + q(3)z^{-3} + ...$$

Therefore, the positive time part of zQ(z) is given by,

$$\left[zQ(z)\right]_+ = \left[z+q(1)+q(2)z^{-1}+q(3)z^{-2}+\ldots\right]_+$$

$$= q(1)+q(2)z^{-1}+q(3)z^{-2}+\ldots$$

$$= z\left[Q(z)-1\right] \qquad \ldots(3)$$

Substituting equation (3) into equation (2), the causal linear predictor becomes,

$$H(z) = \frac{1}{Q(z)} z\left[Q(z)-1\right] = z\left[1-\frac{1}{Q(z)}\right] \qquad \ldots(4)$$

Finally, for the minimum mean square error, we have,

$$\xi_{min} = \frac{1}{2\pi j}\oint_C \left[P_d(z) - H(z)P_{dx}^*(1/z^*)\right]z^{-1}dz \qquad \ldots(5)$$

Since, $P_d(z) = P_x(z)$ and $P_{dx}(z) = ZP_x(z)$, it follows that the minimum mean square error is given by,

$$\xi_{min} = \frac{1}{2\pi j}\oint_C \left[P_x(z) - z^{-1}H(z)P_x^*(1/z^*)\right]z^{-1}dz$$

Using the symmetry of the power spectrum, $P_x(z) = P_x{}^*(1/z^*)$, this becomes,

$$\xi\min = \frac{1}{2\pi j}\oint_C P_x(z)\left[1-z^{-1}H(z)\right]z^{-1}dz$$

Substituting the system function for the causal Wiener linear predictor, equation (4) into this expression yields,

$$\xi_{min} = \frac{1}{2\pi j}\oint_C P_x(z)\left[1-\left(1-\frac{1}{Q(z)}\right)\right]z^{-1}dz$$

$$= \frac{1}{2\pi j}\oint_C \frac{P_x(z)}{Q(z)} z^{-1}dz = \frac{1}{2\pi j}\oint_C \sigma_0^2 Q^*(1/z^*)z^{-1}dz$$

$$= \sigma_0^2 q(0) \qquad \ldots(6)$$

Thus, since Q(z) is monic with q(0) = 1, we have the simple result,

$$\xi_{min} = \sigma_0^2 \qquad \ldots(7)$$

Specifically,

$$\xi_{min} = \exp -\frac{1}{2\pi}\int_{-\pi}^{\pi} \text{In } P_x\left(e^{j\omega}\right)d\omega \qquad ...(8)$$

which is known as the Kolmogorov-Szego formula.

Let us now apply these results to the linear prediction of an auto-regressive process. If x(n) is an AR(p) process with a power spectrum, then,

$$P_x(z) = \frac{\sigma_0^2}{A(z)A^*(1/z^*)}$$

Where,

$$A(z) = 1 + \sum_{k=1}^{p} a(k)z^{-k}$$

is a minimum phase polynomial having all of its roots inside the unit circle, then the optimum linear predictor is given by,

$$H(z) = z\left[1 - A(z)\right] = -a(1) - a(2)z^{-1} - ... - a(p)z^{-p+1} \qquad ...(9)$$

which is an FIR filter.

Therefore, given the entire past history of x(n), only the last p values of x(n) are used in the prediction of x(n + 1). However, an AR(p) random process satisfies a difference equation of the form,

$$x(n) = -a(1)x(n-1) - a(2)x(n-2) - ... - a(p)x(n-p) + w(n)$$

Where, w(n) is the white noise.

Since w(n + 1) cannot be predicted from x(n) or previous values of x(n), then the best that can be done in predicting x(n + 1) is to use the model for x(n) and ignore the noise,

$$\hat{x}(n+1) = -a(1)x(n) - a(2)x(n-1) - \cdots - a(p)x(n-p+1)$$

Of course, this predictor is the same as the one given in equation (9).

2.2.4 Wiener DE Convolution

In mid 1960s, DE convolution or inverse filtering began to be applied broadly to digital image restoration. In most images, adjacent pixels are highly correlated, whereas the gray levels of widely separated pixels are loosely correlated. From this, we can say that the auto-correlation function of typical images generally decreases away from the origin. Since the power spectrum of an image is the Fourier transform of its auto-correlation function, we can say that the power spectrum of an image generally decreases with frequency.

Typical noise sources have either a flat power spectrum or one that decreases with frequency more slowly than typical image power spectra. Thus, the expected situation is for the signal to dominate the spectrum at low frequencies, while the noise dominates at high frequencies. Since the magnitude of the DE convolution filter generally increases with frequency, the filter enhances high frequency noise.

Mean Square Error estimation (MSE) procedure and Wiener DE convolution filter, has the two dimensional transfer function, which is given as,

$$G(u,v)=\frac{H^{*}(u,v)}{|H(u,v)|^{2}+P_{n}(u,v)/P_{f}(u,v)}$$

Where,

p_f = Power spectra of the signal.

P_n = Power spectra of the noise.

Wiener DE convolution affords an optimal method for rolling off the DE convolution transfer function in the presence of noise, but it is plagued with three problems that limit its effectiveness. First, the Mean Square Error criterion of optimality is not particularly good if the image is being restored for the human eye. The problem is that, the MSE criterion weights all errors equally, regardless of their location in the image, whereas the eye is considerably more tolerant of errors in dark areas and high gradient area than elsewhere. In minimizing the Mean Square Error, the Wiener filter also tends to smooth the image more than the eye would prefer.

Second, classical Wiener DE convolution cannot handle a spatially variant blurring PSF. This occurs with coma, astigmatism, curvature of field and with motion blur that involves rotation. Finally, the technique cannot handle the common case of non-stationary signals and noise. Most images are highly non-stationary, having large flat areas separated by sharp transitions. Further, several important noise sources are highly dependent on local gray level. To solve these problems, several alternatives and improvements upon Wiener DE convolution have been developed.

2.2.5 Discrete Kalman Filter

The problem we are seeking to solve is the continual estimation of a set of parameters whose values change over time. Updating is achieved by combining a set of observations or measurements z(t) which contain information about the signal x(t). The role of the estimator is to provide an estimate $\hat{x}(t+\tau)$ at some time $t+\tau$. If $\tau>0$, we have a prediction filter, if $\tau<0$, we have a smoothing filter and if $\tau=0$, then the operation is simply called as filtering.

An estimator is said to be unbiased, if the expectation of its output is the expectation of the quantity being estimated, $E[\hat{x}]=E[x]$.

A Minimum Variance Unbiased Estimator (MVUE) is an estimator which is unbiased and minimizes the mean square error.

$$\hat{x} = \arg\min_{\hat{x}} E\left[\|\hat{x}-x\|^2 |z\right] = E[x|z]$$

The term $E\left[\|x-\hat{x}\|^2\right]$, is called as the variance of error and is closely related to the error covariance matrix, $E\left[(x-\hat{x})(x-\hat{x})^T\right]$. Specifically, the variance of error of an estimator is equal to the trace of the error covariance matrix,

$$E\left[\|x-\hat{x}\|^2\right] = \text{trace } E\left[(x-\hat{x})(x-\hat{x})^T\right]$$

The Kalman filter is a linear minimum variance of error filter over time varying and time invariant filters. In the case of the state vector, x and the observations z being jointly Gaussian distributed, the MVUE estimator is a linear function of the measurement set and thus, the MVUE is also a LMV estimator.

System and Observation Model

We assume that the system can be modelled by the state transition equation,

$$x_{k+1} = F_k x_k + G_k u_k + w_k \qquad ...(1)$$

Where,

F_k = State transition matrix.

x_k = State at time k.

G_k = Input transition matrix.

u_k = Input control vector.

w_k = Additive system or process noise.

We further assume that the observations of the state are made through a measurement system, which can be represented by a linear equation of the form,

$$z_k = H_k x_k + v_k \qquad ...(2)$$

Where,

z_k = Observation or measurement made at time k.

H_k = Observation matrix.

x_k = State at time k.

v_k = Additive measurement noise.

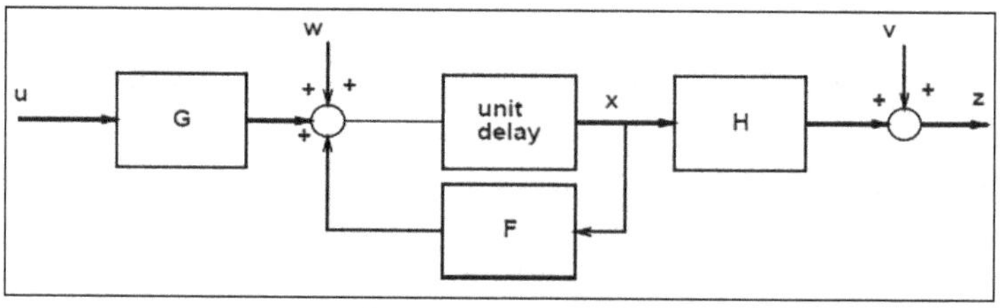

State space model.

Assumptions

We make the following assumptions:

- The process and measurement noise random processes w_k and v_k are uncorrelated, zero mean white noise processes with known covariance matrices. Then,

$$E\left[w_{k.}w_l^T\right] = \begin{Bmatrix} Q_k & k = l, \\ o & \text{otherwise} \end{Bmatrix} \qquad ...(3)$$

$$E\left[v_k v_l^T\right] = \begin{Bmatrix} R_k & k = l, \\ o & \text{otherwise} \end{Bmatrix} \qquad ...(4)$$

$$E\left[w_k v_l^T\right] = o \quad \text{for all } k, l \qquad ...(5)$$

Where, Q_k and R_k are symmetric positive semi-definite matrices.

- The initial system state x_o is a random vector that is uncorrelated to both the system and measurement noise processes.
- The initial system state has a known mean and covariance matrix,

$$\hat{x}_{o|o} = E[x_o] \text{ and } P_{o|o} = E\left[\left(\hat{x}_{o|o} - x_o\right)\left(\hat{x}_{o|o} - x_o\right)^T\right] \qquad ...(6)$$

Given the above assumptions, the task is to determine, given a set of observations $z_1, \ldots, z_{k+1}$, the estimation filter at the k + 1th instance in time generates an optimal estimate of the state x_{k+1}, which we denote by $\hat{x}_{k+1}$, that minimizes the expectation of the squared error loss function,

$$E\left[\left\|x_{k+1} - \hat{x}_{k+1}\right\|^2\right] = E\left[\left(x_{k+1} - \hat{x}_{k+1}\right)^T \left(x_{k+1} - \hat{x}_{k+1}\right)\right] \qquad \ldots(7)$$

Derivation

Consider the estimation of state $\hat{x}_{k+1}$ based on the observation up to time k, $z_1, \ldots, z_k$, namely. This is called as a one step ahead prediction or simply prediction. Now, the solution to the minimization of equation (7) is the expectation of the state at time k + 1 conditioned on the observations up to time k. Thus,

$$\hat{x}_{k+1|k} = E\left[x_{k+1} \mid z_1, \ldots, z_k\right] = E\left[x_{k+1} \mid Z^k\right] \qquad \ldots(8)$$

Then the predicted state is given by,

$$\hat{x}_{k+1|k} = E\left[x_{k+1} \mid Z^k\right]$$

$$= E\left[F_k x_k + G_k u_k + w_k \mid Z^k\right]$$

$$= F_k E\left[x_k \mid Z^k\right] + G_k u_k + E\left[w_k \mid Z^k\right]$$

$$= F_k \hat{x}_{k|k} + G_k u_k \qquad \ldots(9)$$

Where, we have used the fact that, the process noise has zero mean value and u_k is known precisely.

The estimate variance $P_{k+1|k}$ is the mean squared error in the estimate $\hat{x}_{k+1|k}$.

Thus, using the fact that, w_k and $\hat{x}_{k|k}$ are uncorrelated, we have,

$$P_{k+1|k} = E\left[\left(x_{k+1} - \hat{x}_{k+1|k}\right)\left(x_{k+1} - \hat{x}_{k+1|k}\right)^T \mid Z^k\right]$$

$$= F_k E\left[\left(x_k - \hat{x}_{k|k}\right)\left(x_k - \hat{x}_{k|k}\right)^T \mid Z^k\right] F_k^T + E\left[w_k w_k^T\right]$$

$$= F_k P_{k|k} F_k^T + Q_k \qquad \ldots(10)$$

Having obtained a predictive estimate $\hat{x}_{k+1|k}$, suppose that we now take another observation z_{k+1}. We assume that the estimate is a linear weighted sum of the prediction and the new observation and can be described by the equation,

$$\hat{x}_{k+1|k+1} = K'_{k+1}\hat{x}_{k+1|k} + K_{k+1} z_{k+1} \quad \text{...(11)}$$

Where, K'_{k+1} and K_{k+1} are weighting or gain matrices. Our problem is reduced to finding the K_{k+1} and K'_{k+1} that minimize the conditional mean squared estimation error, where the estimation error is given by,

$$\tilde{x}_{k+1|k+1} = \hat{x}_{k+1|k+1} - x_{k+1}$$

2.3 Spectrum Estimation

Many of the important results in Wiener and adaptive filtering assume that the power spectrum or auto-correlation sequence is known. In practice, these quantities need to be estimated. The power spectrum is a quantity which depends on the ensemble characteristics of a stochastic process; it depends on the ensemble of realizations. An estimate of the power spectrum must be made from a finite number of realizations of the process or in the extreme case, a single measured time series of the process.

There are a number of techniques for spectral estimation. These techniques can be grouped into two categories. They are,

- Non-parametric
- Parametric

Non-parametric techniques make no assumption about the underlying model for the stochastic process and instead estimate the power spectrum directly from the observed data. The non-parametric techniques tend to provide a spectral estimate which is not as good as that obtained using parametric techniques, which do make assumptions about the underlying model.

Singular Value Decomposition (SVD) methods have shown great promise in array signal processing. It is assumed that {x(k)} is a WSS stochastic process with auto-correlation sequence $\{r_{xx}(k)\}$, where,

$$r_{xx}(k) = \varepsilon\left[x(n)\times(n+k)\right]$$

Power spectrum is given by,

$$P_{xx}(\theta) = \sum_{k=-\infty}^{\infty} r_{xx}(k)e^{-jk\theta} \quad \text{...(1)}$$

Given a finite collection $\{x_1(k)\},\{x_2(k)\},\ldots,(x_R(k))$ of realizations of {x(k)}, we would like to form an estimate $\hat{P}_{xx}(\theta)$ of the true power spectrum $P_{xx}(\theta)$.

The power spectrum may be expressed in the convenient form as,

$$P_{xx}(\theta)=\lim_{M\to\infty}\varepsilon\left[\frac{1}{2M+1}\left|\sum_{k=-M}^{M}x(k)e^{-jk\theta}\right|^2\right] \qquad \ldots(2)$$

$$\left|\sum_{k=-M}^{M}x(k)e^{-jk\theta}\right|^2=\sum_{k=-M}^{M}\sum_{l=-M}^{M}x(k)x(l)e^{-j(k-l)\theta}$$

That, this may be demonstrated by writing,

Thus,

$$\varepsilon\left[\left|\sum_{k=-M}^{M}x(k)e^{-jk\theta}\right|^2\right]=\sum_{k=-M}^{M}\sum_{l=-M}^{M}r_{xx}(k-l)e^{-j(k-l)\theta}$$

Which we can express equivalently as,

$$\varepsilon\left[\left|\sum_{k=-M}^{M}x(k)e^{-jk\theta}\right|^2\right]=\sum_{m=-2M}^{2M}(2M+1-|m|)r_{xx}(m)e^{-jm\theta} \qquad \ldots(3)$$

Substituting equation (3) into equation (2), we get,

$$\lim_{M\to\infty}\varepsilon\left[\frac{1}{2M+1}\left|\sum_{k=-M}^{M}x(k)e^{-jk\theta}\right|^2\right]=\lim_{M\to\infty}\sum_{m=-2M}^{2M}\left[1-\frac{|m|}{2M+1}\right]r_{xx}(m)e^{-jm\theta} \qquad \ldots(4)$$

If the auto-correlation sequence satisfies,

$$\lim_{M\to\infty}\sum_{m=-2M}^{2M}|m|r_{xx}(m)=0 \qquad \ldots(5)$$

Then equation (2) becomes,

$$\lim_{M\to\infty}\varepsilon\left[\frac{1}{2M+1}\left|\sum_{k=-M}^{M}x(k)e^{-jk\theta}\right|^2\right]=\sum_{m=-\infty}^{\infty}r_{xx}(m)^{-jm\theta}=P_{xx}(\theta)$$

Proving the validity of equation (2).

The condition in equation (5) means that, the auto-correlation sequence has a sufficient rate of decay. This condition is met by stochastic processes with zero mean and no sinusoidal components. If this is not true, the definition of power spectrum can be extended to include Dirac delta functions.

2.3.1 Non-parametric Method: The Periodogram

Non-parametric methods are based on the idea of estimating the auto-correlation sequence of a random process from a set of measured data and then taking the Fourier transform to obtain an estimate of the power spectrum. Although the periodogram is easy to compute, it is limited in its ability to produce an accurate estimate of the power spectrum, particularly for short data records.

The Periodogram

The power spectrum of a wide sense stationary random process is the Fourier transform of the auto-correlation sequence,

$$P_x\left(e^{j\omega}\right) = \sum_{k=-\infty}^{\infty} r_x(k)e^{-jk\omega}$$

Therefore, spectrum estimation is an auto-correlation estimation problem. For an auto-correlation ergodic process and an unlimited amount of data, the auto-correlation sequence may be determined using the time average,

$$r_x(k) = \lim_{x\to\infty} \frac{1}{2N+1} \sum_{n=-N}^{N} x(n+k)x^*(n) \qquad \text{...(1)}$$

However, if x(n) is only measured over a finite interval, say n = 0, 1 N - 1, then the auto-correlation sequence must be estimated using equation (1) with a finite sum,

$$\hat{r}_x(k) = \frac{1}{N}\sum_{n=0}^{N-1} x(n+k)x^*(n) \qquad \text{...(2)}$$

In order to ensure that the values of x(n) that fall outside the interval [0, N - 1] are excluded from the sum.

Equation (2) can be rewritten as follows:

$$\hat{r}_x(k) = \frac{1}{N}\sum_{n=0}^{N-1-k} x(n+k)x^*(n) \; : \; k = 0,1,\dots,N-1 \qquad \text{...(3)}$$

With the values of $\hat{r}_x(k)$ for k < 0 defined using conjugate symmetry, $\hat{r}_x(-k) = \hat{r}_x^*(k)$ and with $\hat{r}_x(k)$ set equal to zero for |k| ≥ N.

Taking the discrete time Fourier transform of $\hat{r}_x(k)$ leads to an estimate of the power spectrum known as the periodogram.

$$\hat{P}_{per}\left(e^{j\omega}\right) = \sum_{k=-N+1}^{N-1} \hat{r}_x(k) e^{-jk\omega} \qquad ...(4)$$

Although defined in terms of the estimated auto-correlation sequence $\hat{r}_x(k)$, it will be more convenient to express the periodogram directly in terms of the process x(n). This may be done as follows:

Let $x_N(n)$ be the finite length signal of length N that is equal to x(n) over the interval [0, N - 1] and is zero otherwise,

$$x_N(n)\begin{Bmatrix} x(n) & ; 0 \le n < N \\ 0 & ; \text{otherwise} \end{Bmatrix} \qquad ...(5)$$

Thus, $x_N(n)$ is the product of x(n) with a rectangular window $w_R(n)$,

$$x_N(n) = w_R(n) \times (n) \qquad ...(6)$$

In terms of $x_N(n)$, the estimated auto-correlation sequence may be written as follows:

$$\hat{r}_x(k) = \frac{1}{N} \sum_{n=-\infty}^{\infty} x_N(n+k) x_N^*(n) = \frac{1}{N} x_N(k) * x_N^*(-k) \qquad ...(7)$$

Taking the Fourier transform and using the convolution theorem, the periodogram becomes,

$$\hat{P}_{per}\left(e^{j\omega}\right) = \frac{1}{N} X_N\left(e^{j\omega}\right) X_N^*\left(e^{j\omega}\right) = \frac{1}{N}\left|X_N\left(e^{j\omega}\right)\right|^2 \qquad ...(8)$$

Where, $X_N\left(e^{j\omega}\right)$ is the discrete time Fourier transform of the N-point data sequence $x_N(n)$,

$$X_N\left(e^{j\omega}\right) = \sum_{n=-\infty}^{\infty} x_N(n) e^{-jn\omega} = \sum_{n=0}^{N-1} x(n) e^{-jn\omega} \qquad ...(9)$$

Thus, the periodogram is proportional to the squared magnitude of the DTFT of $x_N(n)$ and may be computed easily using a DFT as follows:

$$x_N(n) \xrightarrow{DFT} X_N(k) \rightarrow \frac{1}{N}\left|X_N(k)\right|^2 = \hat{P}_{per}\left(e^{j2\pi k/N}\right)$$

2.3.2 Performance of Periodogram

The performance of the periodogram is given with respect to the issues of,

- Spectral Leakage
- Resolution
- Bias of the periodoram
- Variance of the Periodogram

Spectral Leakage: Consider the power spectrum or PSD of a finite length signal $x_L[n]$. It is useful to interpret $x_L[n]$ as the result of multiplying an infinite signal, x[n], by a finite length rectangular window, $w\ [n]$.

$$x_L[n] = x[n]\ .\ w_R[n]$$

Since multiplication in time domain corresponds to convolution in the frequency domain, the Fourier transform of the expression above is given by,

$$X_L(f) = \frac{1}{f_s} \int_{-f_s/2}^{f_s/2} X(\rho) W_R(f-\rho) d\rho$$

The expression for periodogram is given by,

$$\hat{P}_{xx}(f) = \frac{|X_L)f|^2}{f_s L}$$

This illustrates that, the periodogram is also influenced by this convolution.

The effect of the convolution is best understood for sinusoidal data. Suppose that x[n] is composed of a sum of M complex sinusoids, then,

$$x[n] = \sum_{k=1}^{M} A_k e^{j\omega_k n}$$

Its spectrum is given by,

$$X(f) = f_s \sum_{k=1}^{M} A_k \delta(f - f_k$$

Which for a finite length sequence becomes,

$$X_L(f) = \int_{-f_s/2}^{f_s/2} \left(\sum_{k=1}^{M} A_k\ \delta(\rho - f_k) \right) W_R(f-\rho) d\rho = \sum_{k=1}^{M} A_k W_R(f - f_k)$$

So in the spectrum of the finite length signal, the Dirac deltas have been replaced by the terms of the form $W_R(f - f_k)$, which corresponds to the frequency response of a rectangular window centered on the frequency f_k.

The frequency response of a rectangular window has the shape of a sinc signal, as shown below:

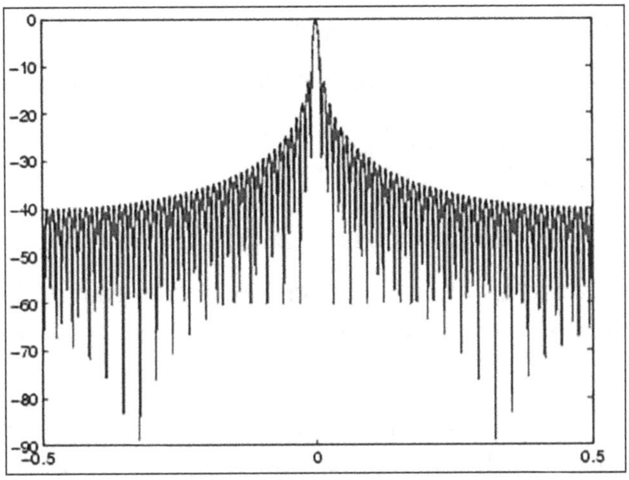

The plot displays a main lobe and several side lobes, the largest of which is approximately 13.5 dB below the main lobe peak. These lobes account for the effect known as the spectral leakage. While the infinite length signal has its power concentrated exactly at the discrete frequency points f_k, the windowed or truncated signal has a continuum of power leaked around the discrete frequency points f_k.

Since the frequency response of a short rectangular window is a much poorer approximation to the Dirac delta function than that of a longer window, spectral leakage is especially evident when data records are short.

Resolution: Resolution refers to the ability to discriminate spectral features and is a key concept on the analysis of spectral estimator performance.

In order to resolve two sinusoids that are relatively close together in frequency, it is necessary for the difference between the two frequencies to be greater than the width of the main lobe of the leaked spectra for either one of these sinusoids. The main lobe width is defined to be the width of the main lobe at the point where the power is half the peak of the main lobe power. This width is approximately equal to f_s/L.

In other words, for two sinusoids of frequencies f_1 and f_2, the resolvability condition requires that,

$$\Delta f = (f_1 - f_2) > \frac{f_s}{L}$$

Bias of the Periodogram: The periodogram is a biased estimator of the PSD. Its expected value can be shown to be,

$$E\left\{\frac{\left|X_L(f)\right|^2}{f_s L}\right\}=\frac{1}{f_s L}\int_{-f_s/2}^{f_s/2} P_{xx}(\rho)\left|W_R(f-\rho)\right|^2 d\rho$$

which is similar to the first expression for $X_L(f)$ in spectral leakage, except that the expression here is in terms of average power rather than magnitude.

This suggests that the estimates produced by the periodogram correspond to a leaky PSD rather than the true PSD.

Note that, $\left|W_R(f-\rho)\right|^2$ essentially yields a triangular Bartlett window. This results in a height for the largest side lobes of the leaky power spectra, that is about 27 dB below the main lobe peak, (i.e.) about twice the frequency separation relative to the non-squared rectangular window.

The periodogram is asymptotically unbiased, which is evident from the observation that, as the data record length tends to ∞, the frequency response of the rectangular window more closely approximates the Dirac delta function.

However, in some cases, the periodogram is a poor estimator of the PSD, even when the data record is long. This is due to the variance of the periodogram.

Variance of the Periodogram: The variance of the periodogram can be shown to be approximately,

$$\operatorname{var}\left\{\frac{\left|X_L(f)\right|^2}{f_s L}\right\}\approx P_{xx}^2(f)\left[1+\left(\frac{\sin(2\pi L f/f_s)}{L\sin(2\pi f/f_s)}\right)^2\right]$$

which indicates that, the variance does not tend to 0, as the data length L tends to ∞.

In statistical terms, the periodogram is not a consistent estimator of the PSD. The periodogram can be a useful tool for spectral estimation, in situations where the SNR is high and especially if the data record is long.

2.4 Parametric Methods: AR Spectrum Estimation

A different approach to spectrum estimation of a random process is the parametric one, which is based on the assumption that, the process can be described by a parametric model. Based on the model, the spectrum of the process can then be expressed in terms of the parameters of the model.

Thus, the approach consists of the following three steps:

- Selection of an appropriate parametric model.
- Estimation of the model parameters.
- Computation of the spectrum using the so obtained parameters.

The parametric spectrum estimation methods are known as high resolution methods, because they can achieve better resolution than the non-parametric methods.

The most frequently used models are the Auto-Regressive (AR), the Moving Average (MA), the Auto-Regressive Moving Average (ARMA) and the sum of harmonics embedded in noise. With the AR model we assume that, the observed data have been generated by a system, whose input-output difference equation is given by,

$$x[n] = -\sum_{k\ 1} a\ x[n-k] + e[n] \qquad ...(1)$$

Where,

$x[n] =$ Observed output of the system.

$e[n] =$ Unobserved input of the system.

$a_k =$ Coefficients.

The input e[n] is a zero means white noise process with unknown variance σ^2 and p is the order of the system. This model is usually abbreviated as AR(p).

The MA model is given by,

$$x[n] = \sum_{k=0}^{q} b_k e[n-k] \qquad ...(2)$$

Where,

$b_k =$ MA parameters.

$e[n] =$ Zero mean white noise process with unknown variance σ^2.

$q =$ Order of the model.

The first MA coefficient b_o is usually set to be 1 and the model is denoted by MA(q).

The ARMA model combines the AR and MA models and is described by,

$$x[n] = -\sum_{k=1}^{p} a_k x[n-k] + \sum_{k=0}^{q} b_k e[n-k] \qquad ...(3)$$

Since the AR and MA orders are p and q, respectively, the model in equation (3) is referred to as ARMA (p, q).

Finally, the model of complex sinusoids in noise is given by,

$$x[n] = \sum_{i=1}^{m} A_i e^{j2\pi f_i n} + e[n], \quad n = 0, 1, ..., N-1 \qquad ...(4)$$

Where,

m = Number of complex sinusoids.

A_i = Complex amplitude of the ith complex sinusoid.

f_i = Complex frequency of the ith complex sinusoid.

$e[n]$ = Sample of a noise process, which is not necessarily white.

Frequently, we assume that the samples e[n] are generated by a certain parametric probability distribution, whose parameters are unknown or e[n] itself is modeled as an AR, MA or ARMA process.

Spectrum Estimation Based on Auto-Regressive Models

When the model of x[n] is AR(p), the PSD of the process is given by,

$$P_{AR}(f) = \frac{\sigma^2}{\left|1 + \sum_{k=1}^{p} a_k e^{-j2\pi fk}\right|^2} \qquad ...(5)$$

Thus, to find PAR(f) we need the estimates of the AR coefficients and the noise variance σ^2.

If we multiply the two sides of equation (1) by x*[n - k], k ≥ 0 and take their expectations, we get,

$$E(x[n]x^*[n-k]) = -\sum_{l=1}^{p} a_l E(x[n-l]x^*[n-k]) + E(e[n]x^*[n-k]) \qquad ...(6)$$

Or,

$$r[k] = \begin{cases} -\sum_{l=1}^{p} a_l r[k-l], & k > 0 \\ -\sum_{l=1}^{p} a_l r[k-l], + \sigma^2, & k = 0 \end{cases} \qquad ...(7)$$

The expressions in equation (7) are known as the Yule-Walker equations. To estimate the p unknown AR coefficients from equation (7), we need at least p equations as well as the estimates of the appropriate auto-correlations.

The set of equations that require the estimation of the minimum number of correlation lags are given by,

$$\hat{R}a = -\hat{r} \qquad ...(8)$$

Where, $\hat{R}$ is the p x p matrix,

$$\hat{R} = \begin{bmatrix} \hat{r}[0] & \hat{r}[-1] & \hat{r}[-2] & \cdots & \hat{r}[-p+1] \\ \hat{r}[1] & \hat{r}[0] & \hat{r}[-1] & \cdots & \hat{r}[-p+2] \\ \vdots & \vdots & \vdots & \vdots & \vdots \\ \hat{r}[p-1] & \hat{r}[p-2] & \hat{r}[p-3] & \cdots & \hat{r}[0] \end{bmatrix} \qquad ...(9)$$

And,

$$\hat{r} = \left[\hat{r}[1]\hat{r}[2]\cdots\hat{r}[p]\right]^T \qquad ...(10)$$

The parameters are estimated by,

$$\hat{a} = -\hat{R}^{-1}\hat{r} \qquad ...(11)$$

The noise variance is found from,

$$\hat{\sigma}^2 = \hat{r}[0] + \sum_{k=1}^{p} a_k \hat{r}^*[k] \qquad ...(12)$$

The PSD estimate is obtained when $\hat{a}$ and $\hat{\sigma}^2$ are substituted in equation (5). This approach for estimating the AR parameters is known as the auto-correlation method.

Many other AR estimation procedures have been proposed including the maximum likelihood method, the covariance method and the Burg method. Burg showed that the AR model provides an extrapolation of a known auto-correlation sequence r[k], $|k| \leq p$, for |k| beyond p, so that, the spectrum corresponding to the extrapolated sequence is the flattest of all spectra consistent with the 2p + 1 known auto-correlations.

An important issue in finding the AR PSD is the order of the assumed AR model. There exist several model order selection procedures, but the most widely used are the Information Criterion A (AIC) and the Information Criterion B (BIC), also known as the Minimum Description Length (MDL) principle. According to the AIC criterion, the best model is the one that minimizes the function AIC(k) over k defined by,

$$AIC(k) = N \log \hat{\sigma}_k^2 + 2k \qquad ...(13)$$

Where,

k = Model order.

$\hat{\sigma}_k^2$ = Estimated noise variance of that model.

Similarly, the MDL criterion chooses the order, which minimizes the function MDL(k) defined by,

$$\mathrm{MDL}(k) = N \log \hat{\sigma}_k^2 + k \log N \qquad ...(14)$$

Where, N is the number of observed data samples.

It is important to emphasize that the MDL rule can be derived if, as a criterion for model selection, we use the Maximum A Posteriori (MAP) principle. It has been found that, the AIC is an inconsistent criterion, whereas the MDL rule is consistent. Here, consistency means that the probability of choosing the correct model order tends to one as, $N \rightarrow \infty$.

The AR based spectrum estimation methods show very good performance, if the processes are narrow band and have sharp peaks in their spectra. Also, many good results have been reported when they are applied to short data records.

2.4.1 MA Spectrum Estimation

The PSD of a moving average process is given by,

$$P_{MA}(f) = \sigma^2 \left| 1 + \sum_{k=1}^{q} b_k e^{-j2\pi fk} \right|^2 \qquad ...(1)$$

It is easy to show that the r[k]'s for $|k| > q$ of an MA(q) process are identically equal to zero and that equation (1) can be expressed as,

$$P_{MA}(f) = \sum_{k=-q}^{q} r[k] e^{-j2\pi fk} \qquad ...(2)$$

Thus, to find $\hat{P}_{MA}(f)$, it would be sufficient to estimate the auto-correlations r[k] and use the found estimates in equation (2). Obviously, this estimate would be identical to $\hat{P}_{BT}(f)$, when the applied window is rectangular and of length 2q + 1.

A different approach is to find the estimates of the unknown MA coefficients and σ^2 and use them in equation (1). The equations of the MA coefficients are non-linear, which makes their estimation difficult. First the data are modeled by an AR model of order L, where L >> q. Then the sequence 1, $\hat{a}_1$, $\hat{a}_2$, ..., $\hat{a}_L$ is fitted with an AR(q) model, whose parameters are also estimated using the auto-correlation method. The estimated coefficients $\hat{b}_1$, $\hat{b}_2$, ..., $\hat{b}_q$ are substituted in equation (1) together with $\hat{\sigma}^2$.

Good results with MA models are obtained when the PSD of the process is characterized by broad peaks and sharp nulls. The MA models should not be used for the processes with narrow band features.

2.4.2 ARMA Spectrum Estimation

The PSD of a process that is represented by the ARMA model is given by,

$$P_{ARMA}(f) = \sigma^2 \frac{\left|1+\sum_{k=1}^{q} b_k e^{-j2\pi fk}\right|^2}{\left|1+\sum_{k=1}^{p} a_k e^{-j2\pi fk}\right|^2} \qquad ...(1)$$

The ML estimates of the ARMA coefficients are difficult to obtain, so we generally use the methods that yield sub-optimal estimates. For example, we can first estimate the AR coefficients based on the equation,

$$\begin{bmatrix} \hat{r}[q] & \hat{r}[q-1] & \cdots & \hat{r}[q-p+1] \\ \hat{r}[q+1] & \hat{r}[q] & \cdots & \hat{r}[q-p+2] \\ \vdots & \vdots & \vdots & \vdots \\ \hat{r}[M-1] & \hat{r}[M-2] & \cdots & \hat{r}[M-p] \end{bmatrix} \begin{bmatrix} a_1 \\ a_2 \\ \vdots \\ a_p \end{bmatrix} + \begin{bmatrix} \epsilon_{q+1} \\ \epsilon_{q+2} \\ \vdots \\ \epsilon_M \end{bmatrix} = - \begin{bmatrix} \hat{r}[q+1] \\ \hat{r}[q+2] \\ \vdots \\ \hat{r}[M] \end{bmatrix} \qquad ...(2)$$

Or,

$$\hat{R}a + \epsilon = -\hat{r} \qquad ...(3)$$

Where, ϵi is a term that models the errors in the Yule-Walker equation due to the estimation errors of the auto-correlation lags and $M \geq p + q$. From equation (3), we can find the least squares estimates of a by,

$$\hat{a} = -\left(\hat{R}^H \hat{R}\right)^{-1} \hat{R}^H \hat{r} \qquad ...(4)$$

This procedure is known as the least square modified Yule-Walker equation method. Once the AR coefficients are estimated, we can filter the observed data and obtain a sequence that is approximately modelled by an MA(q) model.

$$y[n] = x[n] + \sum_{k=1}^{p} \hat{a}_k x[n-k] \qquad ...(5)$$

From the data, y[n], we can estimate the MA PSD and obtain the PSD estimate of the data x[n].

$$\hat{P}_{ARMA}(f) = \frac{\hat{P}_{MA}(f)}{\left|1+\sum_{k=1}^{p} \hat{a}_k e^{-j2\pi fk}\right|^2} \qquad ...(6)$$

Or we can estimate the parameters $b_1, b_2, \ldots, b_q$ and σ^2 and then use,

$$\hat{P}_{ARMA}(f) = \hat{\sigma}^2 \frac{\left|1+\sum_{k=1}^{q} \hat{b}_k e^{-j2\pi fk}\right|^2}{\left|1+\sum_{k=1}^{p} \hat{a}_k e^{-j2\pi fk}\right|^2} \qquad \ldots(7)$$

The ARMA model has an advantage over the AR and MA models, because it can better fit the spectra with nulls and peaks. One disadvantage is that, it is more difficult to estimate its parameters than the parameters of the AR and MA models.

2.5 Frequency Estimation: Eigen Decomposition of the Auto-Correlation Matrix

$$x(n)\sum_{i=1}^{p} A_i e^{jn\omega_i} + w(n)$$

The model in which x(n) is a sum of complex exponentials in white noise is given by,

It is assumed that the amplitudes Ai are complex,

$$A_i = |A_i| e^{j\phi_i}$$

With ϕ_i uncorrelated random variables that are uniformly distributed over the interval $[-\pi, \pi]$.

Although, the frequencies and magnitudes of the complex exponentials, ω_i and $|A_i|$, respectively, are not random, they are assumed to be unknown. Thus, the power spectrum of x(n) consists of a set of p impulses of area $2\pi|A_i|$ at frequency ω_i for i = 1, 2, ..., p, plus the power spectrum of the additive noise w(n). Signals of this form are found in a number of applications such as sonar signal processing and speech processing.

Typically, the complex exponentials are the information bearing part of the signal and it is the estimation of the frequencies and amplitudes that is of interest, rather than the estimation of the power spectrum itself. For example, in case of sonar signals, the frequencies ω_i may represent bearing or velocity information, whereas for speech signals they would correspond to the formant frequencies.

Although, it is possible to estimate the frequencies of the complex exponentials from the peaks of the spectrum that is estimated using any of the techniques, this approach would not fully exploit the assumed parametric form of the process. Here, the methods are based on an eigen decomposition of the auto-correlation matrix that are divided into two sub-spaces, a signal sub-space and a noise subspace. Once these sub-spaces have been determined, a frequency estimation function is then used to extract the estimates of the frequencies.

Eigen Decomposition of the Auto-correlation Matrix

In order to motivate the use of an eigen decomposition of the auto-correlation matrix as an approach that may be used for frequency estimation, we consider the first order process that consists of a single complex exponential in white noise.

$$x(n) = A_1 e^{jn\omega_1} + \omega(n) \qquad \text{...(1)}$$

The amplitude of the complex exponential is, $A_1 = |A_1| e^{j\phi_1}$.

Where,

$\phi_1 =$ Uniformly distributed random variable

$\omega(n) =$ White noise that has a variance of σ_ω^2

The auto-correlation sequence of x(n) is given by,

$$r_x(k) = P_1 e^{jk\omega_1} + \sigma_\omega^2 \delta(k) \qquad \text{...(2)}$$

Where, $P_1 = |A_1|^2$ is the power in the complex exponential. Therefore, the $M \times M$ auto-correlation matrix for x(n) is a sum of the auto-correlation matrix due to the signal, R_s and an auto-correlation matrix due to the noise, R_n.

$$R_x = R_s + R_n \qquad \text{...(3)}$$

It is possible to extract all the parameters of x(n) from the eigenvalues and eigen vectors of R_x as follows:

- Perform an eigen decomposition of the auto-correlation matrix, R_x. The largest eigenvalue will be equal to $MP_1 + \sigma_w{}^2$ and the remaining eigenvalues will be equal to $\sigma_w{}^2$.
- Use the eigenvalues of R_x to solve for the power P_1 and the noise variance as follows:

$$\sigma_w^2 = \lambda_{min}$$

$$P_1 = \frac{1}{M}(\lambda_{max} - \lambda_{min}) \qquad \text{...(4)}$$

- Determine the frequency ω_1 from the eigen vector v_{max} that is associated with the largest eigenvalue using, the second coefficient of v_{max},

$$\omega_i = \arg\{v_{max}(1)\} \qquad ...(5)$$

2.5.1 MUSIC

This Multiple Signal Classification (MUSIC) method was proposed in the late 1970s by Schmidt. The process $\{\bar{x}[n]\}$ is described by m complex sinusoids in white noise. If we form an M x M auto-correlation matrix R, then find its eigenvalues and eigen vectors and rank them, its m eigen vectors corresponding to the m largest eigenvalues span the signal sub-space and the remaining eigen vectors span the noise sub-space.

According to MUSIC, we estimate the noise variance from the M - m smallest eigenvalues of $\hat{R}$,

$$\hat{\sigma}^2 = \frac{1}{M-m}\sum_{i=m+1}^{M} \lambda_i \qquad ...(1)$$

The frequencies from the peak locations of the pseudo spectrum is given by,

$$\hat{P}_{MU}(f) = \frac{1}{\sum_{i=m+1}^{M} \left| e(f)^H v_i \right|^2} \qquad ...(2)$$

Note that there are other ways of estimating the f_i's. Finally the powers of the complex sinusoids are determined.

MUSIC has better performance, because of the introduced averaging via the extra noise eigen vectors. The averaging reduces the statistical fluctuations present in Pisarenko's pseudo spectrum, which arise due to the errors in estimating the auto-correlation matrix. These fluctuations can be reduced further by applying the Eigen vector method, which is a modification of MUSIC and whose pseudo spectrum is given by,

$$\hat{p}_{EV}(f) = \frac{1}{\sum_{i=M+1}^{M} \left| \frac{1}{\lambda_i} e(f)^H v_i \right|^2} \qquad ...(3)$$

Pisarenko's method, MUSIC and its variants exploit the noise sub-space to estimate the unknown parameters of the random process. However, there are approaches that estimate the unknown parameters from vectors that lie in the signal sub-space. The main idea is to form a reduced rank auto-correlation matrix, which is an estimate of the signal auto-correlation matrix.

Since this estimate is formed from the m principal eigen vectors and eigenvalues, the methods based on them are called as principal component spectrum estimation methods. Once the signal auto-correlation matrix is obtained, the frequencies of the complex sinusoids are found, followed by the estimation of the remaining unknown parameters of the model.

3

Adaptive Filtering

3.1 Introduction to Adaptive Filtering

Adaptive linear filters are linear dynamical system with variable or adaptive structure and parameters. They have the property to modify the values of their parameters, (i.e.) their transfer function, during the processing of the input signal, in order to generate signal at the output which is without undesired components, degradation, and noise and interference signals. The goal of adaptation is to adjust the characteristics of the filter through an interaction with the environment in order to reach the desired values.

The operation of adaptive filters is based on the estimation of the statistical properties of the signal in its environment, while modifying the value of its parameters in order to minimize a certain criterion function. The criterion function may be determined in a number of ways, depending on the particular purpose of the adaptive filter, but usually it is a function of some reference signal.

The reference signal may be defined as the desired response of the adaptive filter and in that case the role of the adaptive algorithm is to adjust the parameters of the adaptive filter in such a way to minimize the error signal, which represents the difference between the signal at the output of the adaptive filter and the reference signal.

The basic processes included in adaptive filtering are digital filtering and adaptation or adjustment, (i.e.) the estimation of parameters or coefficients of the filter. The choice of filter structure and the criterion function used during the adaptation process, have the crucial influence to the characteristics of the adaptive filter as a whole.

An adaptive filter is defined by four aspects. They are,

- The signals being processed by the filter.
- The structure that defines how the output signal of the filter is computed from its input signal.
- The parameters within this structure that can be iteratively changed to alter the filter's input- output relationship.
- The adaptive algorithm that describes how the parameters are adjusted from one time instant to the other.

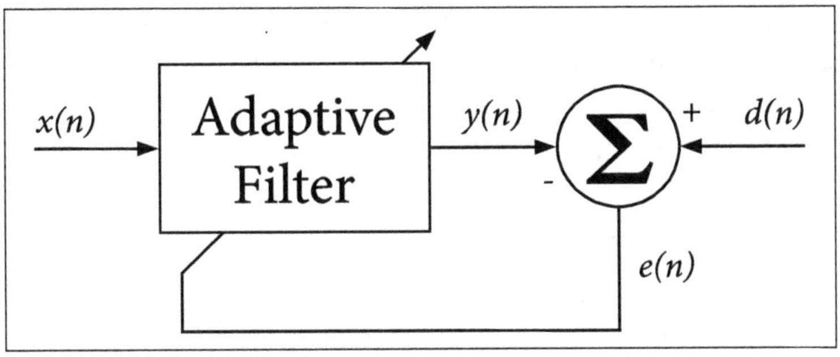

Block diagram of adaptive filtering.

Here,

x(n) = Input digital signal.

y(n) = Output digital signal.

d(n) = Desired response.

e(n) = Error signal.

Problems in Adaptive Filtering:

- The error signal e(n) is calculated from the desired response as shown in the above figure.
- The error signal is fed into a procedure which alters or adapts the parameter of the filter from time n to time (n +1) in a well-defined manner.
- Thus, as time increases the output signal or actual response y(n) is hoped to become better and better and to match the desired response d(n).

An adaptive filter is usually a linear one which can be represented as,

$$y(n)=\sum_{i=0}^{L-1} w_i(n)x(n-i)$$
$$=W^T(n)X(n)$$

Where,

$X(n)=[x(n),\ x(n-1),\ \ldots,\ x(n-L+1)]=$ Input vector.

$W(n)=[w_0(n), w_1(n),\ \ldots,\ w_{L-1}(n)]^T=$ Parameter or co-efficient vector.

3.1.1 FIR Adaptive Filters: The Steepest Descent Adaptive Filter

Filters with adjustable coefficients are called adaptive filters. Such filters incorporate algorithms that allow the filter coefficients to adapt to the signal statistics. The direct form FIR filter structure with adjustable coefficients h(n) is illustrated in figure:

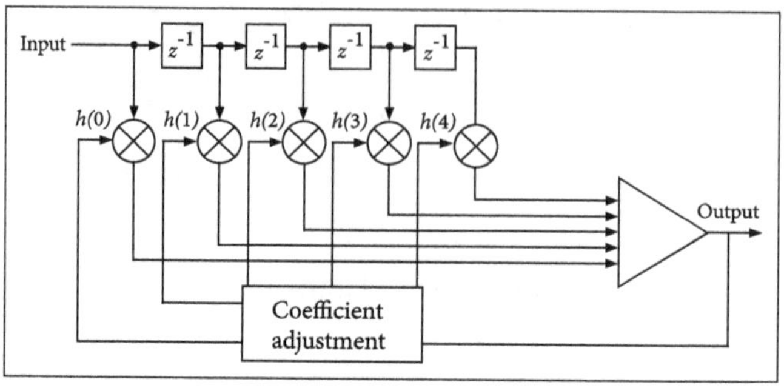

Direct-form Adaptive FIR filter.

System Identification or System Modeling

To identify the unknown system, the system is modeled by an FIR filter with M adjustable coefficients. Both the unknown system and model are excited by an input sequence x(n). If y(n) denotes the output of the unknown system and y^(n) denotes the output of the model,

$$\hat{y}(n)=\sum_{k=0}^{M-1}h(k)x(n-k)$$

From the error sequence e(n) = y(n) – y^(n), n = 0,1,2,....

Select the coefficients h(k) to minimize the error,

$$\xi_M=\sum_{n=0}^{N}\left[y(n)-\sum_{k=0}^{M-1}h(k)x(n-k)\right]^2$$

where N+1 is the number of observations.

The least square criterion leads to the set of linear equations for determining the filter coefficients,

$$\sum_{k=0}^{M-1}h(k)r_{xx}(l-k)=r_{yx}(l),$$

Where, l = 0, 1, 2, ..., M – 1

Where,

$r_{xx}(l)$ is the autocorrelation of the sequence x(n) and $r_{yx}(l)$ is the cross correlation of the system output with the input sequence. By solving the above equation, the filter coefficients are obtained. Since the filter parameters are obtained directly from measurement of data at the input and output of the system without prior knowledge of the plant, the FIR filter model is known as adaptive filter.

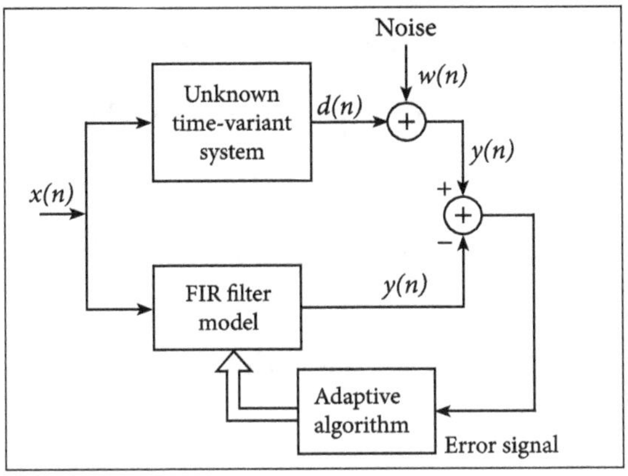

Application of adaptive filtering to system identification.

Applications of Adaptive Filter:

- Adaptive Noise cancellation.
- Echo cancellation in Telephone channels.
- Adaptive Equalization.
- System modeling is used as a model to estimate characteristics of unknown system.
- Adaptive antenna systems to improve beam steering.

Method of Steepest Descent

Given an adaptive filter with an input x(n), an impulse response w(n) and an output y(n) we will get a mathematical relation for the transfer function of the system,

$$y(n) = \underline{w}^T(n)\underline{x}(n)$$

And,

$$x(n) = [x(n),\ x(n-1),\ x(n-2),\ ...,\ x(n-(N-1))]$$

Where, $\underline{w}^T(n) = [w_0(n), w_1(n),\ w_2(n)\ ...\ w_{N-1}(n)] =$ Time domain co-efficient for an Nth order FIR filter.

In the above equation the superscript T represents the transpose of a real valued vector or matrix.

Using an estimate of the ideal cost function the following equation can be derived:

$$w(n+1) = w(n) - \mu\Delta_{E\left[c^2\right]}(n)$$

In the above equation, $\underline{w}(n+1)$ represents the new co-efficient values for the next time interval, μ is a scaling factor and $\Delta E\left[c^2\right](n)$ is the ideal cost function with respect to the vector w(n).

From the above formula we can derive the estimate for the ideal cost function, which is given as,

From the above formula we can derive the estimate for the ideal cost function, which is given as,

$$w(n+1) = w(n) - \mu e(n) x(n)$$

Where,

$$e(n) = d(n) - y(n)$$

And,

$$y(n) = x^T(n) w(n)$$

In the above equation μ is sometimes multiplied by 2, but here we will assume that it is absorbed by a factor μ.

In the Least Mean Squares gradient approximation method, also known as the Method of Steepest Descent, an assumption is made based on the current filter co-efficient and the gradient vector, the derivative of the MSE with respect to the filter co-efficient, is calculated from the assumption. Then an second assumption is made at the tap weight vector by making a change in the present assumption in a direction opposite to the gradient vector. This process is repeated until the derivative of the MSE is zero.

3.2 The LMS Algorithm

The cost function J(n) chosen for the steepest descent algorithm determines the co-efficient solution obtained by the adaptive filter. If the MSE cost function is chosen, the resulting algorithm depends on the statistics of x(n) and d(n), because of the expectation operation that defines this cost function. Since we have only the measurements of d(n) and x(n), we substitute an alternative cost function that depends only on these measurements. One such cost function is the least squares cost function, which is given by,

$$J_{LS}(n)=\sum_{k=0}^{n}\alpha(k)\left(d(k)-W^{T}(n)X(k)\right)^{2} \qquad ...(1)$$

Where, $\alpha(n)$ is a suitable weighting sequence for the terms within the summation. However, this cost function is complicated by the fact that, it requires numerous computations to calculate its value as well as its derivatives with respect to each $w_i(n)$, although efficient recursive methods for its minimization can be developed.

Alternatively, we can propose the simplified cost function $J_{LMS}(n)$ given by,

$$J_{LMS}(n)=\frac{1}{2}e^{2}(n) \qquad ...(2)$$

This cost function can be thought of as an instantaneous estimate of the MSE cost function, as $J_{MSE}(n)=E\{J_{LMS}(n)\}$. Although, it might not appear to be useful, the resulting algorithm obtained when $J_{LMS}(n)$ is used for $J(n)$ is extremely useful for practical applications.

Taking derivatives of $J_{LMS}(n)$ with respect to the elements of $W(n)$, we obtain the LMS adaptive algorithm given by,

$$W(n+1)=W(n)+\mu(n)e(n)X(n) \qquad ...(3)$$

Note that this algorithm is of the general form. It also requires only multiplications and additions to implement. In fact, the number and type of operations needed for the LMS algorithm is nearly the same as that of the FIR filter structure with fixed co-efficient values, which is one of the reasons for the algorithm's popularity.

The behavior of the LMS algorithm has been widely studied and numerous results concerning its adaptation characteristics under different situations have been developed. We indicate its useful behavior by noting that the solution obtained by the LMS algorithm near its convergent point is related to the Wiener solution.

In fact, analyses of the LMS algorithm under certain statistical assumptions about the input and desired response signals show that,

$$\lim_{n\to\infty}E\{W(n)\}=W_{MSE}$$

When the Wiener solution $W_{MSE}(n)$ is a fixed vector.

Moreover, the average behavior of the LMS algorithm is quite similar to that of the steepest descent algorithm that depends explicitly on the statistics of the input and desired response signals. In effect, the iterative nature of the LMS co-efficient updates is a form of time averaging, that smooth's the errors in the instantaneous gradient calculations to obtain a more reasonable estimate of the true gradient.

3.2.1 Convergence of LMS Algorithm

Consider the update equation,

$$\hat{w}(n+1)=\hat{w}(n)+\mu u(n)^{*}e(n)$$

We should know how fast and towards what the algorithm converges, in terms of J(n) and $\hat{w}(n)$.

Strategy

1. Introduce the filter weight error vector $^{o}(n)=\hat{w}(n)-w_0$.
2. Express the update equation in $^{o}(n)$.
3. Express J(n) in $\epsilon(n)$. This leads to $K(n)=E\{\varepsilon(n)\ \varepsilon^{H}(n)\}$.
4. Calculate the difference equation in K(n), which controls the convergence.
5. Make a variable change from K(n) to X(n) which converges equally fast.

1. The filter weight error vector: The LMS contains feedback and just like for the deepest descent there is a risk that the algorithm diverges, if the step size μ is not chosen appropriately.

In order to investigate the stability, the filter weight error vector is introduced $\epsilon(n)$.

$$^{o}(n)=\hat{w}(n)-w_0\left(w_0=R^{-1}p\ \text{Wiener}-\text{Hopf}\right)$$

Note that, $^{o}(n)$ corresponds to $c(n)=w(n)-w_0$ for the steepest descent, but that $^{o}(n)$ is stochastic, because of $\hat{w}(n)$.

2. Update equation in $\epsilon(n)$: The update equation for $^{o}(n)$ at time n + 1 can be expressed recursively as,

$$\begin{aligned}\in(n+1)&=\left[I-\mu\hat{R}(n)\right]\in(n)+\mu\left[\hat{p}(n)-\hat{R}(n)w_0\right]\\&=\left[I-\mu u(n)u^{H}(n)\right]\in(n)+\mu u(n)e_0^{*}(n)\end{aligned}$$

Where, $e_0{}^{*}(n)=d(n)-w_0{}^{H}\,u(n)$

Compare to that of the SD,

$$c(n+1)=(I-\mu R)c(n)$$

3. Express J(n) in ϵ(n): The convergence analysis of the LMS is considerably more complicated than that for the steepest descent.

Therefore, two tools or approximations are needed. They are:

- Independence theory.
- Direct averaging method.

Independence Theory:

- The input vectors of different time instants n, u(1), u(2), ..., u(n), is independent of each other and thereby uncorrelated.
- The input signal vector at time n is independent of the desired signal at all time instants, u(n) independent of d(1), d(2), ..., d(n - 1).
- The desired signal at time n, d(n), is independent of all desired signals, d(1), d(2), ..., d(n - 1), but dependent of the input signal vector at time n, u(n).
- The signals d(n) and u(n) at the same time instant n is mutually normally distributed.

Direct Averaging Method:

Direct averaging means that in the update equation for ϵ,

$$ {}^{o}(n+1)=\left[I-\mu u(n)u^{H}(n)\right]\,{}^{o}(n)+\mu u(n)\,e_{o}{}^{*}(n) $$

The instantaneous estimate $\hat{R}(n)=u(n)u^{H}(n)$ is substituted with the expectation over the ensemble, $R=E\{u(n)\,u^{H}(n)\}$, such that, ${}^{o}(n+1)=(I-\mu R){}^{o}(n)+\mu u(n)e_{o}{}^{*}(n)$ results.

For the SD we could show that,

$$ J(n)=J_{min}+\left[w(n)-w_{o}\right]^{H}R\left[w(n)-w_{o}\right] $$

And that the eigenvalues of R controls the convergence speed.

For the LMS, ${}^{o}(n)=\hat{w}(n)-w_{o}$

$$ J(n)=E\left\{\left|e(n)\right|^{2}\right\}=E\left\{\left|d(n)-\hat{w}^{H}(n)u(n)\right|^{2}\right\} $$

$$ =E\left\{\left|e_{o}(n)-\varepsilon^{H}(n)\,u(n)\right|^{2}\right\} $$

$$\overset{i}{=}\underbrace{E\left\{\left|e_o(n)\right|^2\right\}}_{J_{min}}+E\left\{\in^H(n)\underbrace{u(n)u^H(n)}_{\hat{R}(n)}\in(n)\right\}$$

Here, (i) indicates that the independence assumption is used. Since ϵ and $\hat{R}$ are estimates and in addition not independent, the expectation operator cannot just be moved in to $\hat{R}$.

The behaviour of $J_{ex}(n)=J(n)-J_{min}=E\left\{\varepsilon^H(n)\hat{R}\varepsilon(n)\right\}$ determines the convergence.

Since the following holds for the trace of a matrix,

- tr(scalar) = Scalar.
- tr(AB) = tr(BA).
- E{tr(A)} = tr(E{A}).

$J_{ex}(n)$ can be written as,

$$\begin{aligned}J_{ex}(n)&\overset{1,2}{=}E\left\{tr\left[\hat{R}(n)\in(n)\in^H(n)\right]\right\}\\&\overset{3}{=}tr\left[E\left\{\hat{R}(n)\in(n)\in^H(n)\right\}\right]\\&\overset{i}{=}tr\left[E\left\{\hat{R}(n)\right\}E\left\{\in(n)\in^H(n)\right\}\right]\\&=tr\left[RK(n)\right]\end{aligned}$$

Thus, the convergence depends on $K(n)=E\left\{\varepsilon(n)\ \varepsilon^H(n)\right\}$

4. Difference equation of K(n): The update equation for the filter weight error is given by,

$$^{o}(n+1)=\left[I-\mu u(n)u^H(n)\right]\ ^{o}(n)+\mu u(n)\ e_o{}^*(n)$$

The solution to this stochastic difference equation is for small μ close to the solution for,

$$^{o}(n+1)=[I-\mu R)\ ^{o}(n)+\mu u(n)\ e_o{}^*(n)$$

Which is resulting from direct averaging.

This equation is still difficult to solve, but now the behavior of K(n) can be described by the following difference equation:

$$K(n+1)=(I-\mu R)\ K(n)(I-\mu R)+\mu^2 J_{min}R$$

5. Changing of variables from K(n) to X(n): The variable changes,

$$Q^H R Q = \Lambda$$

$$Q^H K(n) Q = X(n)$$

Where, Λ is the diagonal eigenvalue matrix of R and where X(n) is not normally diagonal, yields,

$$J_{ex}(n) = tr(RK(n)) = tr(\Lambda X(n))$$

The convergence can be described by,

$$X(n+1) = (I - \mu\Lambda) X(n) (I - \mu\Lambda) + \mu^2 J_{min} \Lambda$$

$J_{ex}(n)$ depends only on the diagonal elements of X(n), because of the trace tr[ΛX(n)].

Therefore, we can write, $J_{ex}(n) \sum_{i=1}^{M} \lambda_i x_i(n)$

The update of the diagonal elements is controlled by,

$$x_i(n+1) = (1 - \mu\lambda_i)^2 x_i(n) + \mu^2 J_{min} \lambda_i$$

This is an inhomogeneous difference equation. This means that, the solution will contain a transient part and a stationary part that depends on J_{min} and μ.

Therefore, the cost function can be written as,

$$J(n) = J_{min} + J_{ex}(\infty) + J_{trans}(n)$$

Where,

$J_{min} + J_{ex}(\infty) =$ Stationary solution.

$J_{trans}(n) =$ Transient solutions.

Thus, the LMS can reach $J_{min} + J_{ex}(\infty)$.

3.3 NLMS: Noise Cancellation

In the standard LMS algorithm, the correction is proportional to $\mu x(n) e^*(n)$,

$$W(n+1) = w(n) + \mu x(n)\, e^*(n)$$

If x(n) is large, the update suffers from gradient noise amplification. The Normalized LMS (NLMS) algorithm seeks to avoid gradient noise amplification.

The step size is made time varying, μ(n) and optimized to minimize error.

Thus let,

$$W(n+1)=w(n)+½\,\mu(n)[-\nabla(n)]$$

$$=w(n)+\mu(n)\big[p-Rw(n)\big]$$

Choose μ(n), such that the updated w(n + 1) produces the minimum MSE,

$$J(n+1)=E\left\{\left|e(n+1)\right|\right\}$$

Where,

$$e(n+1)=d(n+1)-w^{H}(n+1)x(n+1)$$

Thus, we choose μ(n) such that it minimizes J(n + 1).

The optimal step size, $\mu_0(n)$, will be a function of R and ∇(n). Here, we use instantaneous estimates of these values.

To determine $\mu_o(n)$, expand J(n + 1).

$$J(n+1)=E\{e(n+1)e^{*}(n+1)\}$$

$$=e\{d(n+1)-w^{H}(n+1)x(n+1))\ (d^{*}(n+1)-x^{H}(n+1)w(n+1))\}$$

$$=\sigma_d{}^2-w^{H}\ (n+1)p-p^{H}\ w(n+1)+w^{H}(n+1)Rw(n+1)$$

Now use the fact that, $w(n+1)=w(n)-½\,\mu(n)\nabla(n)$

$$J(n+1)=\sigma_d^2-\left[w(n)-\frac{1}{2}\mu(n)\nabla(n)\right]^{H}p$$

$$-p^{H}\left[w(n)-\frac{1}{2}\mu(n)\nabla(n)\right]$$

$$+\underbrace{\left[w(n)-\frac{1}{2}\mu(n)\nabla(n)\right]^{H}R\left[w(n)-\frac{1}{2}\mu(n)\nabla(n)\right]}_{\substack{=w^{H}(n)Rw(n)-\frac{1}{2}\mu(n)w^{H}(n)R\nabla(n)\\ -\frac{1}{2}\mu(n)\nabla^{H}(n)Rw(n)+\frac{1}{4}\mu^{2}(n)\nabla^{H}(n)R\nabla(n)}}$$

$$J(n+1) = \sigma_d^2 - \left[w(n) - \frac{1}{2}\mu(n)\nabla(n)\right]^H p$$
$$-p^H\left[w(n) - \frac{1}{2}\mu(n)\nabla(n)\right]$$
$$+w^H(n)Rw(n) - \frac{1}{2}\mu(n)w^H(n)R\nabla(n)$$
$$-\frac{1}{2}\mu(n)\nabla^H(n)Rw(n) + \frac{1}{4}\mu^2(n)\nabla^H(n)R\nabla(n)$$

Differentiating with respect to $\mu(n)$, we get,

$$\frac{\partial J(n+1)}{\partial \mu(n)} = \frac{1}{2}\nabla^H(n)p + \frac{1}{2}p^H\nabla(n) - \frac{1}{2}w^H(n)R\nabla(n)$$
$$-\frac{1}{2}\nabla^H(n)Rw(n) + \frac{1}{2}\mu(n)\nabla^H(n)R\nabla(n)$$

Setting equal to 0.

$$\mu_o(n)\nabla^H(n)R\nabla(n) = w^H(n)R\nabla(n) - p^H\nabla(n)$$
$$+\nabla^H(n)Rw(n) - \nabla^H(n)p$$
$$\mu_o(n) = \frac{\left[w^H(n)R - p^H\right]\nabla(n) + \nabla^H(n)\left[Rw(n) - p\right]}{\nabla^H(n)R\nabla(n)}$$
$$= \frac{\left[Rw(n) - p\right]^H\nabla(n) + \nabla^H(n)\left[Rw(n) - p\right]}{\nabla^H(n)R\nabla(n)}$$

$$\mu_o(n) = \frac{\frac{1}{2}\nabla^H(n)\nabla(n) + \frac{1}{2}\nabla^H(n)\nabla(n)}{\nabla^H(n)R\nabla(n)}$$

$$= \frac{\nabla^H(n)\nabla(n)}{\nabla^H(n)R\nabla(n)}$$

Using instantaneous estimates, we get,

$$\hat{R} = x(n)x^H(n)$$
$$\hat{\nabla}(n) = 2\left[x(n)x^H(n)w(n) - x(n)d^*(n)\right]$$
$$= 2\left[x(n)\left(\hat{d}^*(n) - d^*(n)\right)\right]$$
$$= -2x(n)e^*(n)$$

Thus,

$$\mu_o(n) = \frac{4x^H(n)e(n)x(n)e^*(n)}{2x^H(n)e(n)x(n)x^H(n)2x(n)e^*(n)}$$

$$= \frac{|e(n)|^2 |x^H(n)x(n)|}{|e(n)|^2 (x^H(n)x(n))^2}$$

$$= \frac{1}{x^H(n)x(n)} = \frac{1}{\|x(n)\|^2}$$

Thus, the NLMS update is given by,

$$w(n+1) = w(n) + \underbrace{\frac{\tilde{\mu}}{\|x(n)\|^2}}_{\mu(n)} x(n)e^*(n)$$

To avoid problems when $\|x(n)\|^2 \approx 0$, we add an offset,

$$w(n+1) = w(n) + \frac{\tilde{\mu}}{a + \|x(n)\|^2} x(n)e^*(n)$$

When a > 0.

Now, consider the convergence of the NLMS algorithm.

$$w(n+1) = w(n) + \frac{\tilde{\mu}}{\|x(n)\|^2} x(n)e^*(n)$$

Substituting $e(n) = d(n) - w^H(n)x(n)$, we get,

$$w(n+1) = w(n) + \frac{\tilde{\mu}}{\|x(n)\|^2} x(n)\left[d^*(n) - x^H(n)w(n)\right]$$

$$= \left[I - \tilde{\mu}\frac{x(n)x^H(n)}{\|x(n)\|^2}\right] w(n) + \tilde{\mu}\frac{x(n)d^*(n)}{\|x(n)\|^2}$$

Noise Cancellation

Noise cancellation is a scheme of figuring out signals corrupted by interference or additive noise. From the below figure, it is clear that this system is comprised of primary as well as reference input.

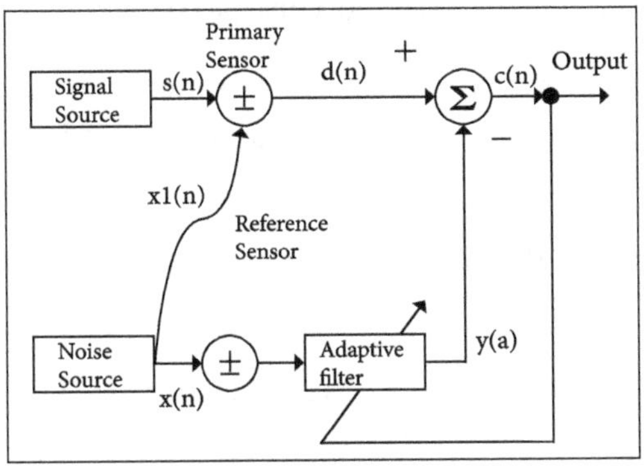

Adaptive noise canceller.

Input to the primary one is a signal source and indirectly a noise source as well, whereas input to the reference is a noise source. The output can be analyzed, which is a combination of signal and noise with adaptive filter.

3.3.1 LMS Based Adaptive Filter

An adaptive filter is a computational device that iteratively models the relationship between the input and output signals of a filter. An adaptive filter self-adjusts the filter coefficients according to the adaptive algorithm. Figure below shows the diagram of a typical adaptive filter.

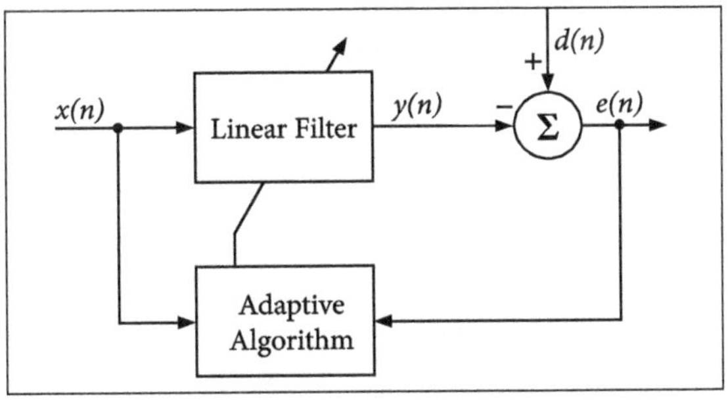

Typical adaptive filter.

Here,

x(n) = Input signal to a linear filter.

y(n) = Corresponding output signal.

d(n) = Additional input signal to the adaptive filter.

e(n) = Error signal that denotes the difference between d(n) and y(n).

A linear filter can be of different filter types such as Finite Impulse Response (FIR) or Infinite Impulse Response (IIR). An adaptive algorithm iteratively adjusts the co-efficient of the linear filter to minimize the power of e(n).

The LMS algorithm is an adaptive algorithm among others, which iteratively adjusts the co-efficient of FIR filters. Other adaptive algorithms include the Recursive Least Square (RLS) algorithms.

The LMS algorithm performs the following operations to update the co-efficients of an adaptive FIR filter:

1. Calculates the output signal, y(n) from the FIR filter.

$$y(n)=\vec{u}^{T}(n)\cdot\vec{w}(n)$$

Where,

$\vec{u}(n)$ = Filter input vector and $\vec{u}(n)=\left[x(n)x(n-1)...x(n-N+1)\right]^{T}$.

$\vec{w}(n)$ = Filter co-efficient vector and $\vec{w}(n)=\left[w_0(n)w_1(n)...w_{N-1}(n)\right]^{T}$.

2. Calculates the error signal, e(n) by using the following equation:

$$e(n)=d(n)-y(n)$$

3. Updates the filter co-efficients by using the following equation:

$$\vec{w}(n+1)=(1-\mu c)\cdot\vec{w}(n)+\mu\cdot e(n)\cdot\vec{u}(n)$$

Where,

μ = Step size of the adaptive filter.

w(n) = Filter co-efficients vector.

u(n) = Filter input vector.

3.3.2 Channel Equalization

Adaptive Channel Equalization

Figure shows a block diagram of a digital communication system in which an adaptive equalizer is used to compensate the distortion caused by the transmission medium. The digital sequence of information a(n) is applied to the transmitting filter whose output is,

$$s(t)=\sum_{k=0}^{\infty}a(k)p(t-kT_s)$$

Where,

$p(t)$ = Impulse response of the filter at transmitter.

T_s = Time interval between information symbols.

Usually linear filter distorts the pulse and causes inter-symbol interference. For example, in telephone channels, filters are used throughout the system which causes phase and amplitude distortion. The distorted signal is corrupted by additive noise.

At the receiver, the signal is first passed through a filter to eliminate the noise outside of the frequency band occupied by the signal.

Samples received at the output of the filter reflect the presence of inter symbol interference and additive noise.

Adaptive equalization is done in two steps. They are:

- Training mode.
- Tracking mode.

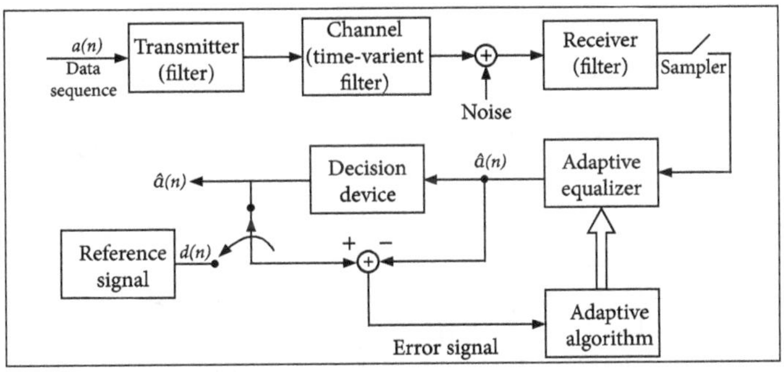

Channel equalization using Adaptive filter.

In training mode, a known test signal is transmitted and that signal is compared in the receiver, resultant error signal gives information about the channel. This error signal is used to adjust the filter coefficients of equalizer. The output of equalizer is connected to decision device to get the estimate of $\hat{a}(n)$.

3.4 Adaptive Recursive Filter

The recursive filter with poles as well as zeros would offer the advantages such as resonance, sharp cut-off, etc. On the other hand, recursive adaptive filters have the following two disadvantages that are not found in the adaptive linear combiner:

- They become unstable, if the poles move outside the unit circle during the adaptive process.

- Generally, their performance surfaces are non-quadratic and may even have local minima.

These are the serious disadvantages and consequently the recursive adaptive filter have very limited applications. Instability must be prevented by limiting the coefficients in some manner and neither the steepest descent method nor Newton's method will work correctly on a multimodal performance surface.

Regarding the second disadvantage, there is some indication that, if the recursive adaptive filter has a sufficient number of zeros and poles, the error surface will be uni-modal. Thus, one may be able to remove the local minima by simply adding the filter weights. To develop algorithms for the recursive adaptive filter, let us place the recursive filter in the standard adaptive mode, as shown in the below figure. Here, the vector X_k can represent either the multiple or single input situation and of course y_k is a scalar.

$$y_k = \sum_{n=0}^{L} a_n x_{k-n} + \sum_{n=1}^{L} b_n y_{k-n} \qquad ...(1)$$

This form applies to the single input case. A time varying weight vector and a new signal vector, W_k and U_k are now defined as follows:

$$W_k = [a_{ok}\ a_{1k} ... a_{Lk}\ b_{1k} ... b_{Lk}]T \qquad ...(2)$$

$$U_k = [x_k\ x_{k-1} ... X_{k-L}\ y_{k-1} ... Y_{k-L}]T \qquad ...(3)$$

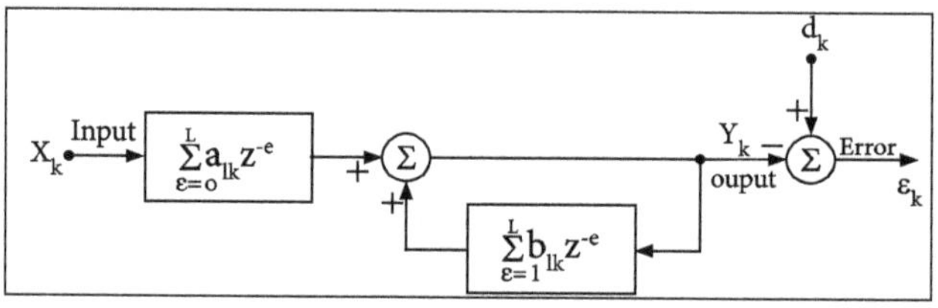

Recursive adaptive filter.

From above figure and from equation (1), we can write,

$$\varepsilon_k = d_k - y_k$$

$$= d_k - W_k^T U_k \qquad ...(4)$$

This is quite similar to the non-recursive case, the main difference is that U_k contains the value of y as well as x.

Considering first the LMS algorithm, we again use the gradient approximation,

$$\hat{\nabla}_k = \frac{\partial \varepsilon^2}{\partial W_k} = 2\varepsilon \frac{\partial \varepsilon}{\partial W_k}$$

$$= 2\varepsilon_k \left[\frac{\partial \varepsilon_k}{\partial a_{ok}} \cdots \frac{\partial \varepsilon_k}{\partial a_{Lk}} \frac{\partial \varepsilon_k}{\partial b_{1k}} \cdots \frac{\partial \varepsilon_k}{\partial b_{Lk}} \right]^T$$

$$= -2\varepsilon_k \left[\frac{\partial y_k}{\partial a_{ok}} \cdots \frac{\partial y_k}{\partial a_{Lk}} \frac{\partial y_k}{\partial b_{1k}} \cdots \frac{\partial y_k}{\partial b_{Lk}} \right]^T \quad \text{...(5)}$$

The derivatives in equation (5) present a special problem, because y_k is now a recursive function. Using equation (1), we define,

$$\alpha_{nk} \triangleq \frac{\partial y_k}{\partial a_n} = x_{k-n} + \sum_{l=L}^{L} b_l \frac{\partial y_{k-1}}{\partial a_n}$$

$$= x_{k-n} + \sum_{l=1}^{L} b_l \alpha_{n,k-1} \quad \text{...(6)}$$

$$\beta_{nk} \triangleq \frac{\partial y_k}{\partial b_n} = y_{k-n} + \sum_{l=L}^{L} b_l \frac{\partial y_{k-1}}{\partial b_n}$$

$$= y_{k-n} + \sum_{l=1}^{L} b_l \beta_{n,k-1} \quad \text{...(7)}$$

With the derivatives defined in this manner, we have,

$$\hat{\nabla}_k = -2\varepsilon_k \left[\alpha_{ok} \quad \cdots \quad \alpha_{Lk}\beta_{1k} \quad \cdots \quad \beta_{Lk} \right]^T \quad \text{...(8)}$$

We write the LMS algorithm as follows:

$$W_{k+1} = W_k - M\hat{\nabla}_k \quad \text{...(9)}$$

Here we have replaced the constant, μ, with the following diagonal matrix:

$$M = \text{diag}\left[\mu \ldots \mu v_1 v_L\right] \quad \text{...(10)}$$

With the non-quadratic error surface, we now have a convergence parameter μ for each a and then a different convergence factor for each b. We may even wish to have these factors vary with time. Using current values of the b's in equation (6) and (7), the LMS algorithm for recursive adaptive filters is as follows:

IIR LMS

$$y_k = W_k^T U_k$$

$$\alpha_{nk} = x_{k-n} + \sum_{l=1}^{L} b_{lk} \alpha_{n,k-1} \quad 0 \le n \le L$$

$$\beta_{nk} = y_{k-n} + \sum_{l=1}^{L} b_{lk}\beta_{n,k-l} \quad 1 \le n \le L$$

$$\hat{\nabla}_k = -2(d_k - y_k)[\alpha_{ok} \quad \cdots \quad \alpha_{Lk}\beta_{1k} \quad \cdots \quad \beta_{Lk}]^T$$

$$W_{k+1} = W_k - M\hat{\nabla}_k \qquad \text{...(11)}$$

The α's and β's should be set initially at zero unless their values are known. Note that the signal vector, U_k, is defined in equation (3) and that b_{lk} is one of the feedback weights of W_k in equation (2).

$$A_k(z) = \sum_{l=o}^{L} a_{lk} z^{-1} \quad \text{and} \quad B_k(z) = \sum_{l=1}^{L} b_{lk} z^{-1} \qquad \text{...(12)}$$

We can write the transfer function for either the second line or third line of equation (11) as,

$$\text{transfer function} = \frac{z^{-n}}{1 - B_k(z)}$$

For example, the computation of α_{nk} is shown in the below figure:

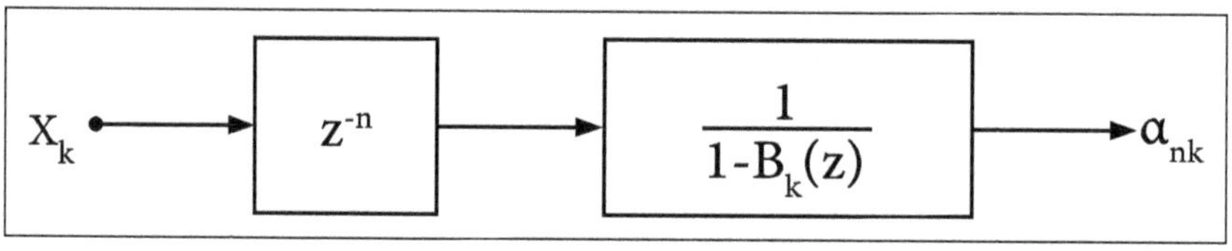

Configuration for producing α_{nk}.

Using this result, the entire adaptive filter is shown in the below figure.

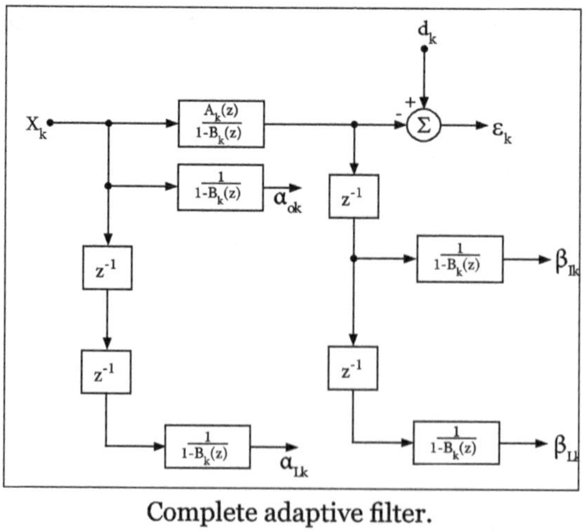

Complete adaptive filter.

3.4.1 Exponentially Weighted RLS

Recursive Least Squares Estimation

Given the set of input samples {u(1), u(2), ..., u(N)} and the set of desired response {d(1), d(2), ..., d(N)}.

- In the family of linear filters, compute their output according to,

$$y(n)=\sum_{k=0}^{M} w_k u(n-k), \quad n=0,1,2, \qquad ...(1)$$

Find recursively in time the parameters $\{w_0(n), w_1(n), ..., w_{M-1}(n)\}$ so as to minimize the sum of error squares.

$$\varepsilon(n)=\varepsilon(w_0(n), w_1(n), ..., w_{M-1}(n))$$

$$=\sum_{i=i_1}^{n}\beta(n,i)\left[e(i)^2\right]\sum_{i=i_1}^{n}\beta(n,i)\left[d(i)-\sum_{k=0}^{M-1}w_k(n)u(i-k)\right]^2$$

Where, the error signal is given by,

$$e(i)=d(i)-y(i)=d(i)-\sum_{k=0}^{M-1}w_k(n)u(i-k)$$

The forgetting factor or weighting factor reduces the influence of old data.

$$0<\beta(n, i)\leq 1, \text{ where, } i = 1, 2, ..., n$$

Usually it takes the form, $(0 < \lambda < 1)$

$$\beta(n, i)=\lambda^{n-i}, \text{ where, } i=1, 2, ..., n$$

Exponentially Weighted Least Squares Solution

Writing the criterion with an exponential forgetting factor gives,

$$\varepsilon(n)=\varepsilon(w_0(n), w_1(n), ..., w_{M-1}(n))$$

$$=\sum_{i=i_1}^{n}\lambda^{n-1}\left[e(i)^2\right]=\sum_{i=i_1}^{n}\lambda^{n-i}\left[d(i)-\sum_{k=0}^{M-1}w_k(n)u(i-k)\right]^2$$

Make the following variable changes:

$$u'(i)=\sqrt{\lambda^{n-i}}u(i); \quad d'(i)=\sqrt{\lambda^{n-i}}d(i) \qquad ...(2)$$

Then the criterion can be rewritten as,

$$\varepsilon(n)=\sum_{i=i_1}^{n}\lambda^{n-i}\left[d(i)-\sum_{k=0}^{M-1}w_k(n)u(i-k)\right]^2=\sum_{i=i_1}^{n}\left[d'(i)-\sum_{k=0}^{M-1}w_k(n)u'(i-k)\right]^2$$

Which is the standard LS criterion, in the new variables u'(i) and d'(i).

The LS solution can be written as,

$$\underline{w}(n)=\sum_{i-i_1}^{n}\underline{u'}(i)\underline{u'}(i)^T\Big)^{-1}\sum_{i=i_1}^{n}\underline{u'}(i)d'(i)$$

$$=\left(\sum_{i=i_1}^{n}\lambda^{n-i}\underline{u}(i)\underline{u}(i)^T\right)^{-1}\sum_{i=i_1}^{n}\lambda^{n-i}\underline{u}(i)d(i)=\left[\Phi(n)\right]^{-1}\underline{\psi}(n)\Big]$$

Where, we will denote,

$$\Phi(n)=\sum_{i\ 1}\lambda^{n\ i}\underline{u}(i)\underline{u}(i)$$

$$\underline{\psi}(n)=\sum_{i=1}^{n}\lambda^{n-i}\underline{u}(i)d(i)$$

3.4.2 Sliding Window RLS

For adaptive purposes, we need an effective window of data.

1. Sliding window:

ϕ_n and θ_n are based only on the last L samples.

$$\phi_{n+1}=\phi_n+x^*(n+1)x^T(n+1)-x^*(n-L)x^T(n-L)$$

$$\theta_{n+1}=\theta_n+x^*(n+1)d(n+1)-x^*(n-L)d(n-L)$$

This doubles complexity and we have to keep L previous data samples in memory.

2. Exponential window:

Scale down ϕ_n and θ_n by a factor $\lambda\approx 1$.

$$\phi_{n\ \ 1}=\lambda\phi_n+x^*(n+1)x\ \ (n+1)$$

$$\theta_{n+1} = \lambda\theta_n + x^*(n+1)d(n+1)$$

This corresponds to,

$$X_{n+1} = \left[x(n+1) \quad \lambda^{1/2}x(n) \quad \lambda x(n-1) \quad \lambda^{3/2}x(n-2) \quad \cdots\right]$$
$$d_{k+1} = \left[d(n+1) \quad \lambda^{1/2}d(n) \quad \lambda d(n-1) \quad \lambda^{3/2}d(n-2) \quad \cdots\right]$$

4

Cardiovascular System

4.1 Cardiovascular System: Heart Structure

The cardiovascular system is responsible for transporting nutrients and removing gaseous waste from the body. This system is comprised of the heart and the circulatory system. Structures of the cardiovascular system include the heart, blood vessels and blood. The lymphatic system is also closely associated with the cardiovascular system.

Structure of the Cardiovascular System

Heart

The heart is the organ that supplies oxygen and blood to all parts of the body. This muscle produces electrical impulses through a process known as cardiac conduction. These impulses cause the heart to contract and then relax, producing what is known as the heartbeat. The beating of heart drives the cardiac cycle which pumps the blood to cells and tissues of the body.

Blood Vessels

Blood vessels are intricate networks of hollow tubes that transport blood throughout the entire body. Blood travels from the heart through arteries to smaller arterioles, then to capillaries or sinusoids, to venules, to veins and back to the heart. Through the process of microcirculation, substances such as carbon dioxide, oxygen, nutrients and wastes are exchanged between the blood and the fluid that surrounds the cell.

Blood

The blood delivers nutrients to the cell and removes wastes that are produced during cellular processes, such as cellular respiration. Blood is composed of white blood cells, red blood cells, platelets and plasma.

Red blood cells contain enormous amount of protein called hemoglobin. This iron containing molecule binds oxygen, as oxygen molecules enter blood vessels in the lungs and transport them to various parts of the body. After depositing oxygen to cells and tissue, red blood cells pick up carbon dioxide for transportation to the lungs, where CO_2 is expelled from the body.

Circulatory System

The circulatory system supplies the body's tissues with oxygen rich blood and important nutrients. In addition to removing gaseous waste, like CO_2, the circulatory system also transports blood to organs such as the liver and kidneys to remove other harmful substances. This system aids in cell to cell communication and homeostasis by transporting hormones and signal messages between the different cells and organ systems of the body.

The circulatory system also transports blood along pulmonary and systemic circuits. The pulmonary circuit involves the path of circulation between the heart and the lungs. The systemic circuit involves the path of circulation between the heart and the rest of the body. The aorta distributes oxygen rich blood to various regions of the body.

Lymphatic System

The lymphatic system is a component of the immune system and works closely with the cardiovascular system. The lymphatic system is a vascular network of tubules and ducts that collect, filter and return lymph to blood circulation.

Lymph is a clear fluid that comes from blood plasma, which exits blood vessels at capillary beds. This fluid becomes the interstitial fluid that bathes tissues and helps to deliver nutrients and oxygen to cells. In addition to returning lymph to circulation, lymphatic structures also filter blood of microorganisms, such as bacteria and viruses.

Lymphatic structures also remove cancerous cells, cellular debris and waste from the blood. Once filtered, the blood is returned to the circulatory system.

Location, Size and Shape of the Heart

The heart is located underneath the sternum in a thoracic compartment called the mediastinum, which occupies the space between the lungs. It is approximately the size of a man's fist, (i.e.) 250-350 grams and is shaped like an inverted cone.

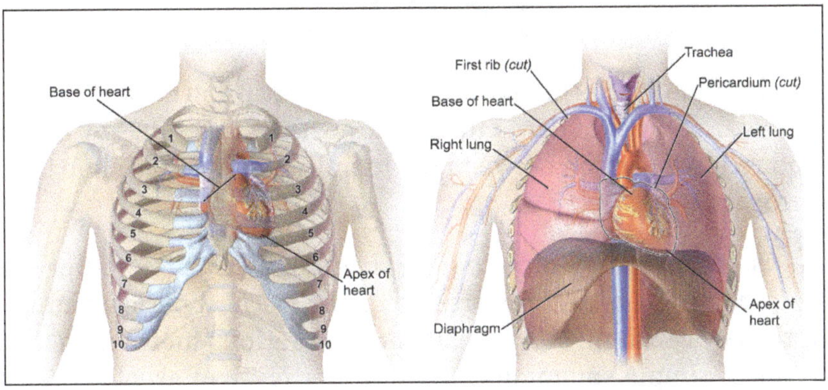

The narrow end of the heart is known as the apex. It is directed downward and to the left and lie just above the arch of the diaphragm at the approximate level of the fifth or sixth rib. The broad end of the heart is known as the base and gives rise to major blood vessels, which is directed upwards and to the right and lies at the approximate level of the second rib.

Surrounding the hearts is a fibrous sac known as the pericardium, which performs several functions. Fluid within the sac lubricates the outer wall of the heart, so it can beat without causing friction. It also holds the heart in place forms a barrier against infections and helps keep the heart from over expanding.

The pericardium is made up of a coronal section which comprises of two walls and a thin intervening space. The outer wall is the thickest and consists of two tissue layers. The external layer is formed by a dense irregular connective tissue and is often referred to as the fibrous pericardium. This layer protects the heart and anchors it to the nearby organs.

At the roots of the major blood vessels, the parietal pericardium reflects back over the surface of the heart to form the inner wall of the pericardium, the visceral pericardium, because it is the outer layer of the heart wall, the visceral pericardium is known as the epicardium. Together, the parietal and visceral paricardial layers are also called as the serous pericardium.

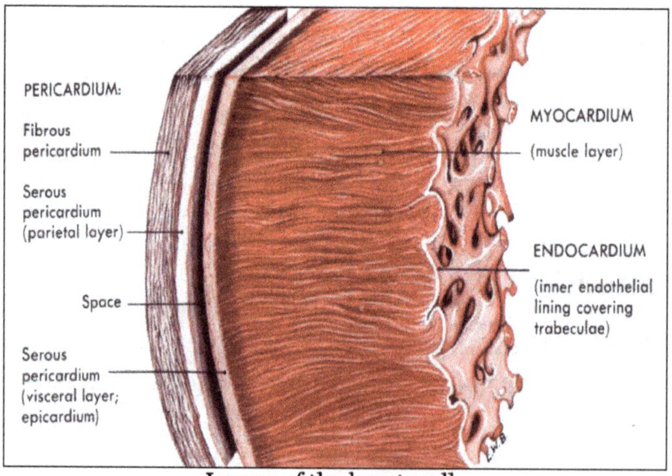

Layers of the heart wall.

Between the walls of the serous pericardium is the pericardial cavity. This narrow space is filled with a few milliliters of pericardial fluid, which is secreted by the serous membrane. The fluid reduces friction between the membranes, as they glide past one another during heartbeats.

The heart wall is composed of three tissue layers. Covering the outer surface of the heart is the epicardium. It is also referred to as the visceral pericardium, which is the inner layer of the pericardium. The epicardium is a serious membrane that consists of an external layer of simple squamous and an inner layer of areolar tissue. The squamous cells secrete lubricating fluids into the pericardial cavity.

The thick middle layer of the heart wall is known as the myocardium. It consists of numerous layers of cardiac muscle fibers that wrap around the heart wall. Contraction of the myocardium pumps blood out of the heart into the aorta and pulmonary trunk arteries. Covering the outer surface of the heart wall is the endocardium. This layer also covers the heart valves and tendons and is continuous with the endothelium that lines the major blood vessels that attach to the heart.

The endocardium is made up of thin layer of simple squamous cells and areolar tissue, similar to the epicardium. Secretions from the squamous cells help regulate the activity of the myocardium.

4.1.1 Cardiac Cycle

The cardiac cycle is the sequence of events that occur when the heart beats. The cycle has two main phases: diastole, when the heart ventricles are relaxed and systole, when the ventricles contract. In cardiac cycle, blood enters the right atrium of the heart from the superior and inferior vena cava and flows across the tricuspid valve into the right ventricle. From the right ventricle, the blood flows into the pulmonary artery, which is separated from the ventricle by the pulmonary valve.

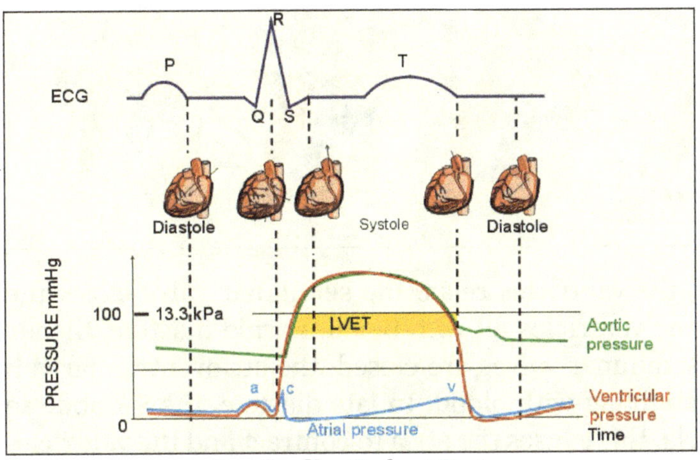

One Cardiac Cycle at Rest.

After oxygenation in lungs, blood returns to the heart via four pulmonary veins that enter the left atrium. From the left atrium, blood flows across the mitral valve and into the left ventricle. From the left ventricle blood is ejected across the aortic valve into the aorta. Together, the mitral and tricuspid valves are known as the atrioventricular valves and the aortic and pulmonary valves are known as the semilunar valves.

The cardiac cycle is due to the movement of blood as a result of the pressure differences within the chambers of the heart. In order for blood to flow through a blood vessel or through a heart valve, there must be a force acting on the blood. This force is provided by the difference in blood pressure across these structures by the contractions of the heart.

Each heart beat or cardiac cycle, is divided into two phases, contraction and relaxation, stimulated by electrical impulses from the Sinoatrial node (SA node). The time during which ventricular contraction occurs is called as the systole. The time between ventricular contraction, during which ventricular filling occurs is called as the diastole, also known as the relaxation phase.

Structure	Phase	Atrial systole	Early ventricular systole ↔ Late ventricular systole	Early ventricular ↔ Late ventricular diastole
Atria		Contract	Relax	Relax
Ventricles		Relax	Contract	Relax
AV valves		Open	Closed	Open
Semilunar valves		Closed	Open	Closed

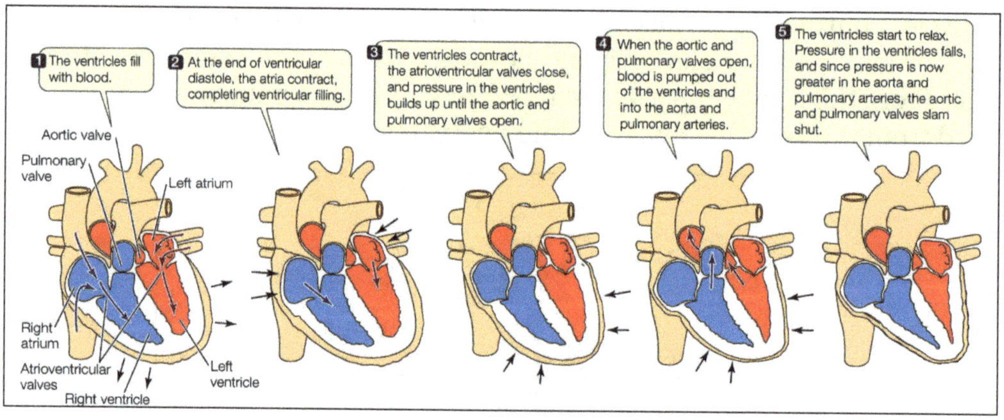

In early diastole, the ventricles relax, the semilunar valve close, the atrioventricular valves open and the ventricles fill with blood. In mid diastole, the atria and ventricles are relaxed, the semilunar valves are closed, the atrioventricular valves are open and the ventricles keep filling with blood. In late diastole, the SA node sends an electrical impulse to the atria, this causes the atria to contract and the ventricles to fill with more blood.

The electrical signal that causes contraction moves from the atria toward the ventricles. Before it does, it reaches the Atrioventricular node (AV node). The AV node delays the signal, so that the ventricles can contract all at once rather than a little bit at a time.

Prior to systole, the electrical signal passes from the AV node down the AV bundle, also known as the bundle of His to the Purkinje fibers. The fibers allow the fast spread of the electrical signals to all parts of the ventricles and the electrical signal causes the ventricles to contract. Systole begins with the closure of the atrioventricular valves. During systole, the ventricles contract, the semi-lunar valves open and blood is pumped from the ventricles to the aorta.

Blood pressure is highest during systole and lowest during diastole. It has two components, the systole and diastole pressure. Normal systole pressure for an adult is estimated at 120 mm hg and normal diastole pressure is estimated at 80 mm hg.

4.1.2 Electrocardiogram (ECG)

Principle

An electrocardiogram (ECG) sometimes known as the EKG is a reading of the electrical activity of the heart. The potentials originated in the individual fibers of heart muscle are added to produce the ECG waveform; the ECG reflects the rhythmic electrical depolarization and repolarization of the myocardium, (i.e.) heart muscle, associated with the contractions of the atria and ventricles.

There are electrical signals that can be detected from two different types of cardiac muscle, the autorhythmic fibers that make up the electrical conduction system of the heart and the cardiac muscles that produce muscle contractions. The signals are detected by electrodes that measure electrical potential between different points on the body.

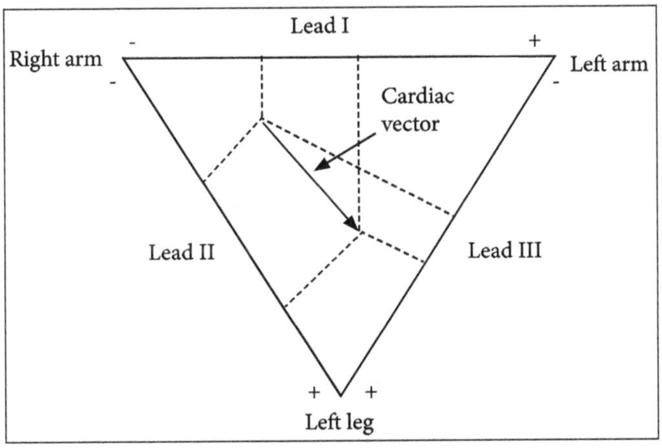

Einthoven triangle.

The vector sum of the frontal plane cardiac vector at any instant onto the three axes of the Einthoven Triangle will be zero.

- Lead I: Potential between the Right Arm (RA) and the Left Arm (LA).
- Lead II: Potential between the Right Arm and the Left Leg.
- Lead III: Potential between the Left Arm and the Left Leg.

A typical ECG is taken using 10 wires and is referred to as a 12 lead ECG. Electrodes are placed on each limb and are referred to as the augmented limb leads and 6 precordial leads are placed directly on the chest. The final lead is the ground lead and is generally placed on the right leg, although it can be placed anywhere on the body.

The ground is used as a point of reference. The leads will be placed on the inside of the forearms near the elbow joint and the ground lead will be placed on the right wrist.

Standard ECG

The electrical conduction system of the heart consists of the Sinoatrial node, located in the superior wall of the right atrium, which is the heart's primary pacemaker. These fibers automatically produce an electrical signal at a particular pace that is conveyed to the Atrioventricular node, located in the medial wall of the right atrium.

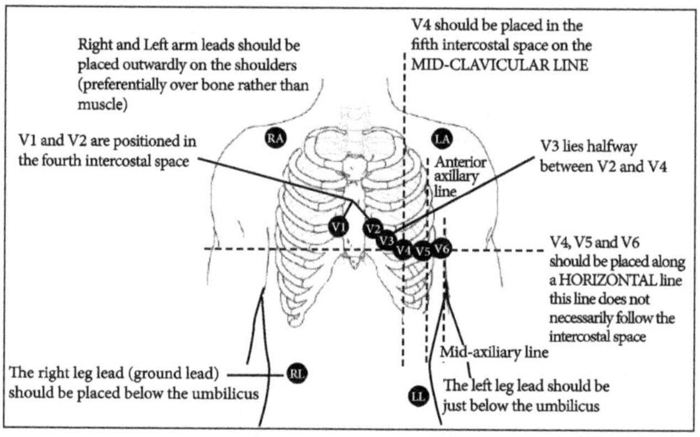

12 lead ECG electrode placement.

The signal from the AV node travels to the atrioventricular bundle and the right and left bundle branches, located in the interventricular septum. From the septum, the signal travels to the Purkinje fibers, which project along the outer walls of the ventricles.

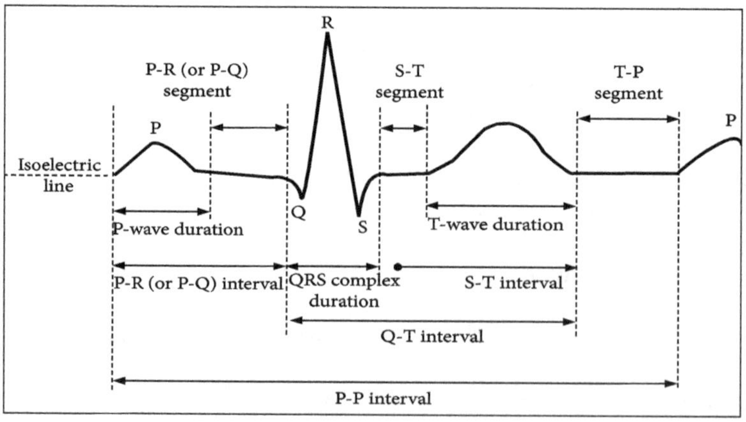

Not only does the signal travel through the autorhythmic fibers that make up the electrical conduction system, but the electrical signal travels through gap junctions to the cardiac muscles, where electrical impulses lead to cardiac muscle contractions. The ECG is the electrical activity of the heart as it is displayed on an oscilloscope. It has five different wave deflections: P, Q, R, S and T.

In general, ECGs are useful for detecting the abnormal rhythms caused by damage to the autorhythmic fibers or abnormal levels of electrolytes such as potassium, sodium or calcium. They are not as useful in detecting damage to cardiac muscle.

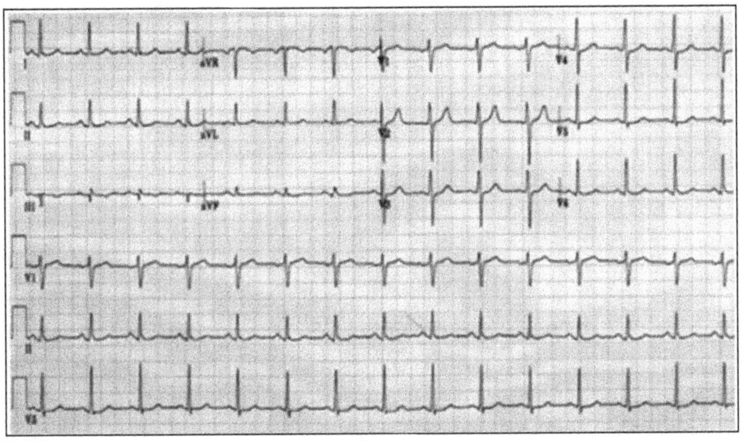

The P, Q, R, S and T deflections and intervals between indicate the following:

- The P wave represents the electrical signal produced by the SA node in the right atrium and the propagation of that impulse to the AV node and to the cardiac muscles of the atria.
- The PR/PQ interval runs from the beginning of the P wave to the beginning of the QRS complex. Atrial contraction occurs during the flat part of the interval.
- The QRS complex represents the depolarization of the ventricles. Reflected in this is the movement of the electrical signal down the AV bundle, right and left bundle branches, Purkinje fibers and the propagation of that signal to the cardiac muscles of the ventricles, because of the greater muscle mass of the ventricles, the signal is much larger than the P wave.

Also during this time we have repolarization of the atrial cardiac muscles, however the electrical signal is masked by the large depolarization produced by the ventricular muscles.

- The ST interval represents the contraction of the ventricles. It is measured from the junction of where QRS ends and the T wave begins.
- The T wave represents the repolarization of the ventricles. The absolute refractory period occurs during the ST interval and extends to the apex of the T wave.
- The QT interval runs from the beginning of the QRS complex to the end of the T wave. This represents the time it takes for depolarization and repolarization of the ventricles. This interval will vary with heart rate.

Typical Times for Segments of an ECG

Segment	Time (ms)
P/Q interval	120-200
QRS complex	60-100
ST interval	80-120
Q/T interval	300-440

Bipolar Limb Leads

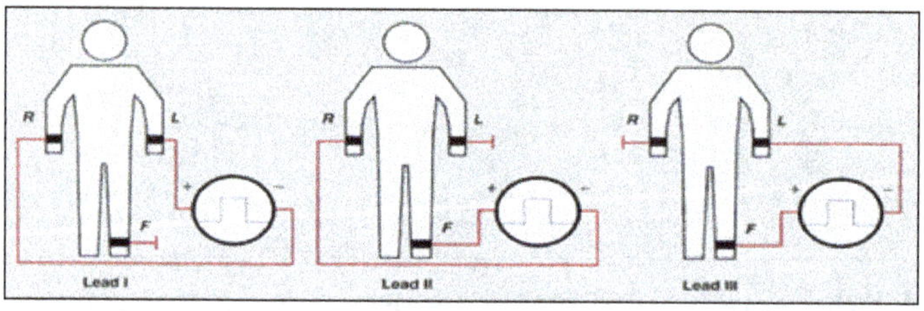

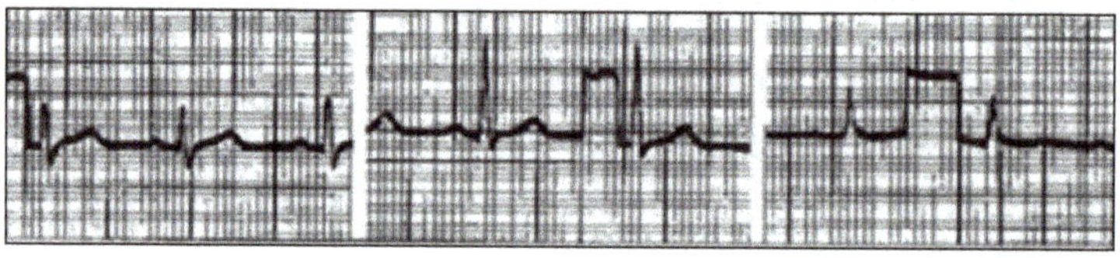

Unipolar Limb Leads

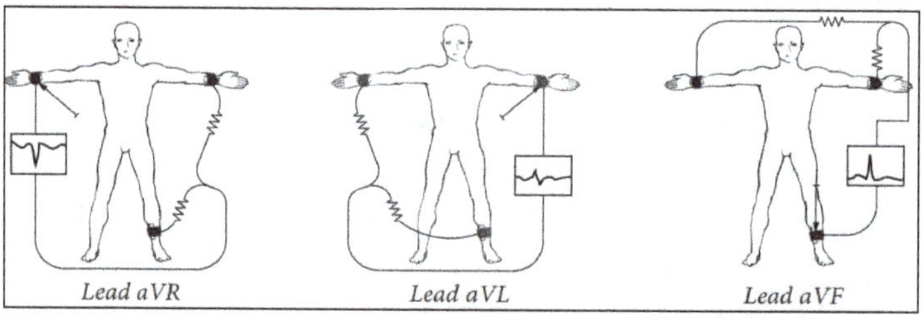

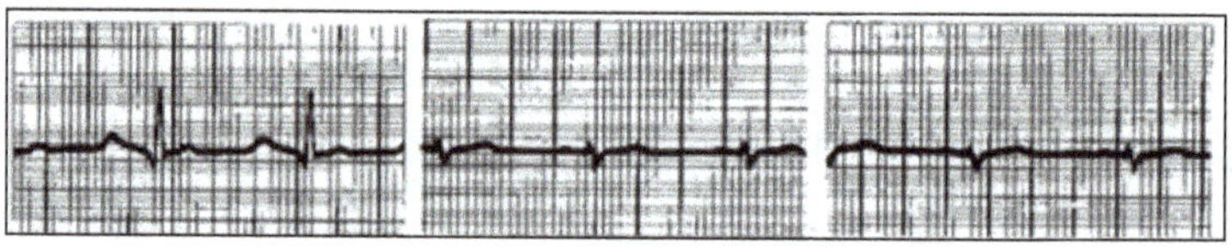

Unipolar Chest Leads

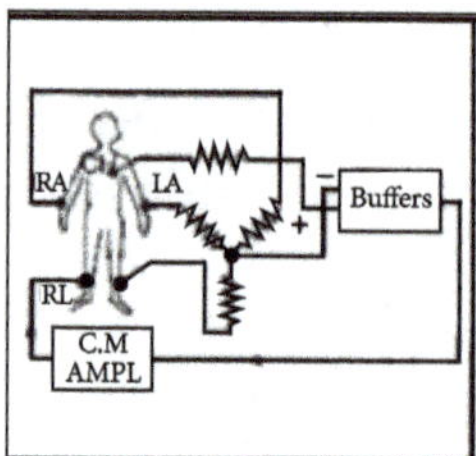

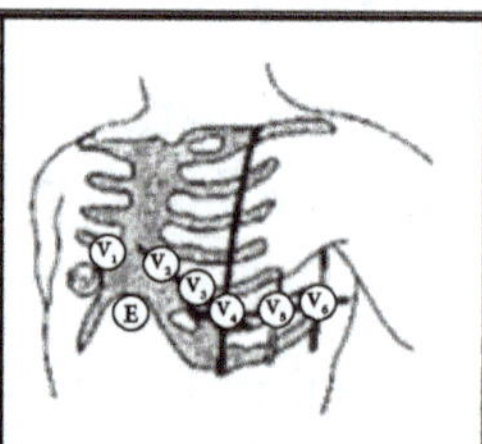

V_1 Fourth intercostal space, at right sternal margin.

V_2 Fourth intercostal space, at left sternal margin.

V_3 Midway between V_2 and V_4

V_4 Fith intercostal space, at mid-calvicular line.

V_5 Same level as V_4 on anterior axillary line.

V_6 Same level as V_4, on mid avitary line.

E Ensiform, base of stomum.

Lead V ++ | CH positions | CH positions

4.1.3 Phonocardiogram (PCG)

Phonocardiography (PCG) is a useful technique for registering heart sounds and murmurs. It was developed in the early days to overcome the deficiencies of the acoustical stethoscope properties. A phonocardiogram, which is the graphic representation of the heart sounds and murmurs, can document the timings and annotates the different relative intensities of heart sounds and murmurs.

Cardiologists can then evaluate a phonocardiogram based on the changes in the wave shape, (i.e.) morphology and the timing parameters. A number of other heart related variables may also be recorded simultaneously with the phonocardiogram, including the ECG, carotid arterial pulse, jugular venous pulse and apex cardiogram. Such information allows clinicians to evaluate the heart sounds of a patient with respect to the electrical and mechanical events of the cardiac cycle.

Although phonocardiography can record and store auscultatory findings accurately, its usage as a diagnostic tool is quite uncommon, because of the required critical procedures and complicated instrumentation. A standard procedure to record the phonocardiograms require a specially designed, acoustically quiet room. In terms of equipment, phonocardiographic devices were introduced before compact analog integrated circuits were available, they are typically large, noisy and inconvenient to use.

With the introduction of the electronic stethoscope, phonocardiography may possibly make a comeback in clinical practice. The newly developed electronic stethoscopes are more compact, robust to noise, immune to disturbance and much more convenient for diagnostic use. They permit a digital recording of heart sounds and therefore, allow clinicians to perform analysis of the phonocardiogram on a PC platform.

The phonocardiogram may have even greater diagnostic importance in the future as further improvements are made in the electronic measurement technology and signal processing algorithms. These algorithms can help to represent the heart sound as a set of characteristic signal components, which may form the detection basis of various heart diseases. Much of the research effort has been devoted to the exploration of signal processing methods with reduced sensitivity to the recorded noise and improved identification of the exact start and end points of major heart sounds.

In general, three ways to represent and characterize heart sound exist. They are,

- Spectrum estimation.
- Time-frequency analysis.
- Non-linear analysis.

The most popular method to represent heart sound components in the past was spectrum estimation. A variety of techniques have been developed to analyze the frequency contents of heart sound. Earlier studies used spectral estimation techniques to extract features in the frequency domain. Classical methods for Power Spectral Density (PSD) estimation employ various windowing and averaging functions to improve the statistical stability of the spectrum obtained by Fast Fourier Transform (FFT).

Early researches have demonstrated some success in distinguishing normal from abnormal patients based on the average power spectrum of diastolic heart sounds, estimated by traditional FFT methods. Despite the success, classical methods may not provide an accurate power spectrum when the signal-to-noise ratio (SNR) is low and the length of the desired signal is short.

An alternative spectral estimation technique is the parametric modelling method, example, Auto-Regressive (AR), Moving Average (MA) and Auto-Regressive Moving Average (ARMA). Parametric modelling involves choosing an appropriate model for the signal and estimating the model parameters. These parameters can then be used to characterize the power spectrum of the signal, to classify the signal or to perform data compression and pattern recognition tasks. Among these modelling techniques, AR modelling shows outstanding performance when the signal has very sharp peaks. The ARMA model can be used to model signals with sharp frequency peaks and valleys.

The application of parametric modelling methods to signal identification can provide a good estimation of spectral features, particularly for a signal with low SNR, which leads to the use of model based methods in the analysis of heart sounds and the detection of features associated with coronary artery diseases.

Another class of spectral estimation techniques is non-parametric Eigen-analysis based method, which is based on Eigen decomposition of the data or its autocorrelation matrix. The eigenvector method, also known as the minimum-norm method can be used to extract the signal buried in noise.

In theory, the eigenvector method has infinite resolution and provides an accurate spectral estimation regardless of the SNR. The eigenvector method has demonstrated the best diagnostic performance when compared with the other spectral estimation methods like FFT, AR and ARMA. In general, the parametric spectral estimators and the eigenvector method offers the promise of higher resolution over the FFT.

However, the major shortcomings of AR, ARMA and eigenvector methods are that, in each case, poor spectral estimation occurs, if the assumed model is inappropriate or if the model orders chosen are incorrect. Early methods of frequency domain analysis were mostly based on the direct applications of Fourier transform or auto-regressive spectral estimation techniques.

However, if the statistical properties of the signal change with time, the direct application of these techniques may be inappropriate, since important time events like frequency variation would be lost in the transformation process. As heart sounds are non-stationary in nature, recent research adopts the time-frequency analysis in order to capture the temporal variation in the heart sound signal.

4.1.4 Electroencephalogram (EEG)

An electroencephalogram (EEG) is a test used to evaluate the electrical activity in the brain. Brain cells communicate with each other through electrical impulses. An EEG can be used to help detect potential problems associated with this activity.

The test tracks and records brain wave patterns. Small, flat metal discs called electrodes are attached to the scalp with wires. The electrodes analyze the electrical impulses in the brain and send signals to the computer, where the results are recorded.

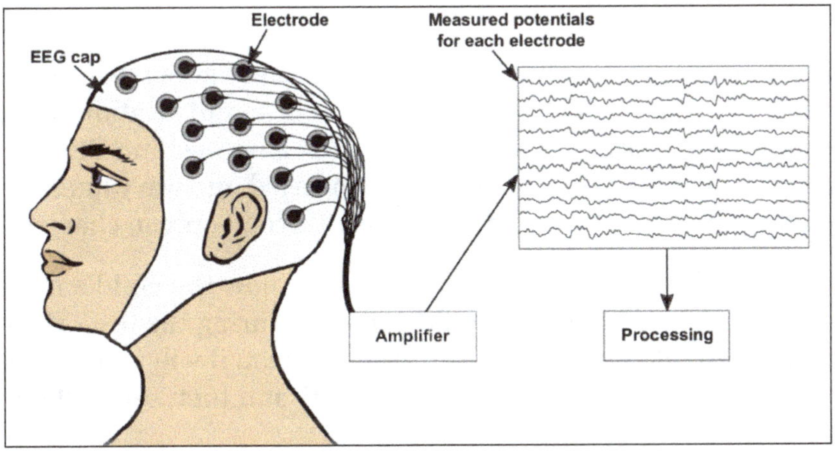

The electrical impulses in an EEG recording look like wavy lines with peaks and valleys. These lines allow doctors to quickly assess whether there are abnormal patterns. Any irregularities may be a sign of seizures or other brain disorders. An EEG is used to detect problems in the electrical activity of the brain that may be associated with certain brain disorders.

The measurements given by an EEG are used to confirm or rule out various conditions, including,

- Brain tumor.
- Dementia.
- Encephalopathy (a disease that causes brain dysfunction).
- Encephalitis (an inflammation of the brain).
- Head injury.
- Memory problems.

- Sleep disorders.
- Stroke.
- Seizure disorders (such as epilepsy).

When someone is in a coma, an EEG may be performed to determine the level of brain activity. The test can also be used to monitor the activity during brain surgery.

There are no risks associated with an EEG. The test is painless and safe. When someone has epilepsy or another seizure disorder, the stimuli presented during the test may cause a seizure. However, the technician performing the EEG is trained to safely manage the situation.

Procedure

An EEG can be done in the lab or hospital. The child will be asked to lie on a bed or sit in a chair. The EEG technician will attach electrodes to different locations on the scalp using adhesive paste. Each electrode is connected to an amplifier and EEG recording machine. The electrical signals from the brain are converted into wavy lines on a computer screen. The child will be asked to lie still, because movements can change the results.

The health care provider performing the EEG will know the child's medical history and will be ready for any issues that could come up during the test. Most EEGs take about an hour. If the child needs to sleep during the test, it will take longer. A neurologist interprets the recordings taken from the EEG and then sends the result to our doctor.

Normal Results

Electrical activity in the brain is seen in an EEG as a pattern of waves. Different levels of consciousness, such as sleeping and waking, have a specific range of frequencies of waves per second and are considered as normal. For example, the wave patterns move faster when we are awake than when we are asleep. The EEG will show if the frequency of waves or patterns are normal. Normal activity means that we don't have a brain disorder.

4.2 X-ray

An X-ray is a common imaging test that is been used for decades. X-rays are a form of electromagnetic radiation, such as the radio waves, infrared radiation, visible light, ultraviolet radiation and microwaves. It can help our doctor view the inner parts of our body without having to make an incision. This can help them diagnose, monitor and treat many medical conditions.

Different types of X-rays are used for different purposes. For example, our doctor may order an X-ray with a barium enema to get a closer look of our gastrointestinal tract. There are some risks involved in getting an X-ray, but for most people, the potential benefits outweigh the risks.

One of the most common and beneficial uses of X-rays are for medical imaging. It is also used in treating cancer and exploring the cosmos.

X-ray Diffraction

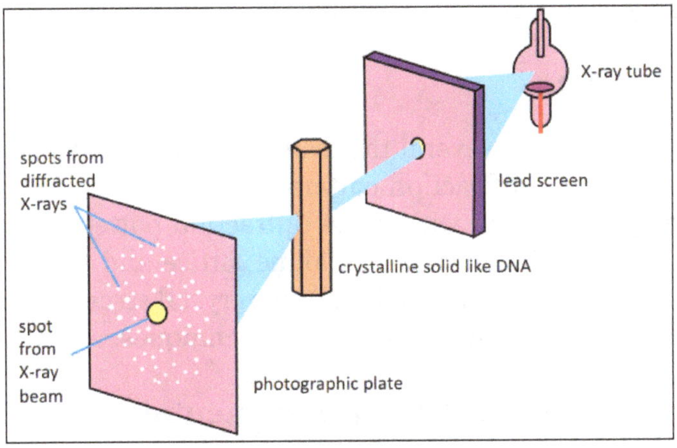

X-ray Diffraction Technique.

Single-crystal X-ray Diffraction is a non-destructive analytical technique which provides detailed information about the internal lattice of crystalline substances, including unit cell dimensions, bond-lengths, bond-angles and details of site-ordering. Directly related is single-crystal refinement, where the data generated from the X-ray analysis is interpreted and refined to obtain the crystal structure.

Fundamental Principles of Single-crystal X-ray Diffraction

X-ray diffraction is based on constructive interference of monochromatic X-rays and a crystalline sample. These X-rays are generated by a cathode ray tube, filtered to produce monochromatic radiation, collimated to concentrate and directed toward the sample.

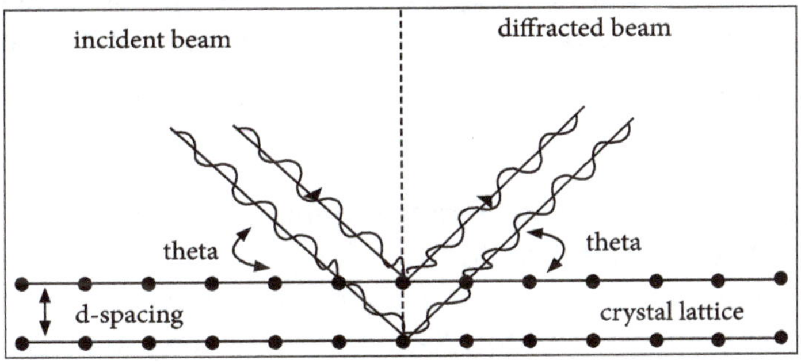

X-ray Diffraction principle.

The interaction of incident rays with the sample produces the constructive interference when conditions satisfy Bragg's Law $(n\lambda = 2d \sin\theta)$. This law relates the wavelength of the electromagnetic radiation to the diffraction angle and the lattice spacing in a crystalline sample. These diffracted X-rays are then detected, processed and then counted. By changing the geometry of the incident rays orientation of the centered crystal and the detector, all possible diffraction directions of the lattice should be attained.

All diffraction methods are based on generation of X-rays in an X-ray tube. These X-rays are directed at the sample and the diffracted rays are collected. A key component of all diffraction is the angle between the incident and the diffracted rays.

Interpretation of Data

Typical mineral structures has several thousand unique reflections, whose spatial arrangement is referred as the diffraction pattern. Indices (hkl) may be assigned to each reflection, indicating its position within the diffraction pattern. It has the reciprocal Fourier transform relationship to the crystalline lattice and the unit cell in the real space. This step is referred to as the solution of the crystal structure. After the structure is solved, it is further refined using least-squares techniques.

Single-crystal X-ray Diffraction: Instrumentation

X-ray diffractometers consist of three basic elements such as an X-ray tube, a sample holder and an X-ray detector. X-rays are generated in a cathode ray tube by heating a filament to produce electrons, which is accelerating the electrons toward a target by applying a voltage and the impact of the electrons with the target material.

When electrons have sufficient energy to dislodge inner shell electrons of the target material, characteristic X-ray spectra are produced. These spectra consist of several components, the most common being Kα and Kβ. Kα consists, in part, of $K\alpha_1$ and $K\alpha_2$. $K\alpha_1$ has a slightly shorter wavelength and also twice the intensity as $K\alpha_2$.

The specific wavelengths are characteristic of the target material. Filtering, by foils or crystal monochromators, is required to produce monochromatic X-rays needed for diffraction. $K\alpha_1$ and $K\alpha_2$ are sufficiently close in wavelength such that a weighted average of the two is used. Molybdenum is the most common target material used for single-crystal diffraction, with radiation $= 0.7107Å$. These X-rays are collimated and directed onto the sample.

When the geometry of the incident X-rays impinging the sample satisfies the Bragg Equation, constructive interference occurs. The detector records and processes this X-ray signal and it converts the signal to a count rate which is then output to a device such as a printer or computer monitor. X-rays may also be produced using a synchrotron, which emits a much stronger beam.

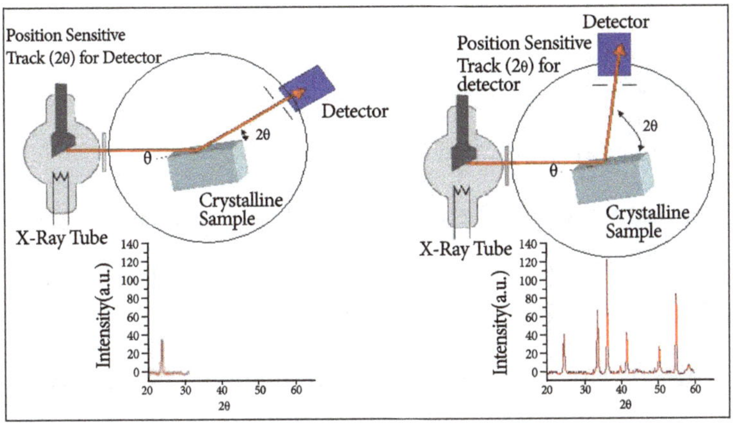

X-ray diffractometer

X-rays leave the collimator and they are directed at the crystal. Rays are either reflected off the surface, transmitted through the crystal or diffracted by the crystal lattice. A beam stop is located directly opposite the collimator to block the transmitted rays and to prevent the burn-out of the detector. Reflected rays are not picked up by the detector due to the angles involved. The diffracted rays at the correct orientation for the configuration are then collected by the detector.

Modern single-crystal diffractometers uses CCD (charge-coupled device) technology to transform the X-ray photons into the electrical signal which are then sent to the computer for processing.

Applications

Single-crystal X-ray diffraction is most commonly used for precise determination of a unit cell, including cell dimensions and positions of atoms within the lattice. Specific applications of single-crystal diffraction include:

- Determination of unit cell, bond-lengths, bond-angles and site-ordering.
- New mineral identification, crystal solution and refinement.
- Variations in crystal lattice with chemistry.
- Characterization of cation-anion coordination.
- Determination of crystal-chemical versus environmental control on mineral chemistry.
- With specialized chambers, structures of high pressure and/or temperature phases can be determined.
- Powder patterns can also be derived from single-crystals by use of specialized cameras.

Advantages

- Non-destructive.
- No separate standards required.
- Determination of crystal-chemical controls on mineral chemistry.
- Detailed crystal structure, including unit cell dimensions, bond-lengths, bond-angles and site-ordering information.
- Powder patterns can also be derived from single-crystals by use of specialized cameras.
- With specialized chambers, structures of high pressure and temperature phases can be determined.

Limitations

- Optically clear sample.
- Must have a single, robust (stable) sample, generally between 50 to 250 microns in size.
- Data collection generally requires between 24 and 72 hours.
- Twinned samples can be handled with difficulty.

4.3 Sonography

Sonography, sometimes called ultrasonography, is the practice of using high frequency waves to produce an image for medical analysis. These high frequency waves are called as ultrasound waves. Ultrasound waves are a useful method to produce a needed medical image without the use of radiation. Sonography is also used to analyze tissues and organs of the body, such as the heart, eyes and abdomen.

Sonogram

Sonogram is an instrument used to visualize the image of internal parts of a body by using high frequency ultrasonic sound waves.

Principle

It uses a probe consisting of transducers to send pulses of sound into a body. Whenever a sound pulse reaches a boundary between two tissue structures, it is partly reflected and partly transmitted. The sound pulse reflected back to the probe is detected as an echo. The reflection depends on the difference between the speeds of the ultrasound in the tissues. The time taken by the echo to travel back to the probe is measured and it is used to determine the depth of tissue interface causing the echo.

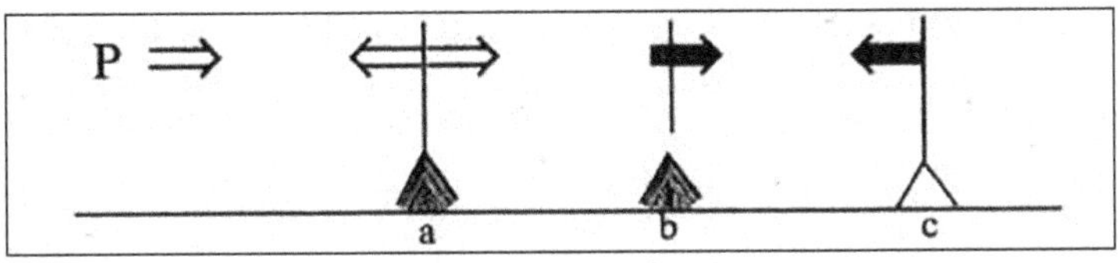

Sonogram echo

Working

- The high frequency ultrasonic waves are transmitted into human body by using transmitting transducer.
- The reflected ultrasonic waves from the internal parts of the body are received by a receiving transducer (piezo electric crystal).
- The electrical signals from the receiver circuit are converted into digital signal by a digital processing unit and they are stored in a memory device.

4.3.1 CT-Scan

A Computerized Tomography (CT) scanner is a special kind of X-ray machine. Instead of sending out a single X-ray through our body as with ordinary X-rays, several beams are sent simultaneously from different angles. This allows more detailed images from within the body to be constructed and these images are then interpreted by a doctor.

CT scanners may also be referred to as CAT scans (Computerized Axial Tomography). Unlike an MRI scanner, where we are placed within a tunnel, a CT scanner consists of a doughnut shaped machine and therefore, we should not feel claustrophobic.

Working

The X-rays from the beams are detected after they have been passed through the body and their strength is measured. Beams that have passed through less dense tissue such as the lungs will be stronger, whereas beams that have passed through denser tissue such as bone will be weaker.

A computer can use this information to work out the relative density of the tissues examined. Each set of measurements made by the scanner is, in effect, a cross-section through the body. The computer processes the results, displaying them as a two-dimensional picture shown on the monitor. The technique of CT scanning was developed by the British inventor Sir Godfrey Hounsfield, who was awarded the Nobel Prize for his work.

Uses

- CT scans are far more detailed than ordinary X-rays.
- The CT scanner was originally designed to take pictures of the brain. Now it is much more advanced and is used for taking pictures of virtually any part of the body.
- The information from the two-dimensional computer images can be reconstructed to produce three-dimensional images by some modern CT scanners.
- CT is used for the detection, diagnosis and treatment of a number of vascular diseases, which may ultimately lead to stroke, kidney failure or blood clots in the lungs.
- This also includes Abdominal Aortic Aneurysm (AAAs), which is where the large main artery running downwards in the abdomen becomes enlarged and therefore, may be prone to spontaneous rupture.
- CT is also used for diagnosing and analyzing many spinal problems and injuries which may occur to the hands, feet and other skeletal structures.
- This is because CT is good, not just for looking at soft tissue structures, but also in providing detailed images of even very small bones.
- They can also be used to produce virtual images that show what a surgeon would see during an operation. They are used for a variety of different reasons, mainly because CT imaging is one of the best and fastest tools for examining the chest and abdomen, because it can provide cross-sectional views and highly detailed images.
- Some of the most common uses of CT imaging include detecting different types of cancer, examining patients with severe injuries and finding the cause for sudden rapid onset symptoms such as breathlessness, abdominal pain, etc.
- CT scans also allow doctors to inspect the inside of the body without having to operate or perform unpleasant examinations.
- It also allows surgeons to plan surgery prior to starting a procedure and allows assessment of the results after a procedure has been performed.
- CT scanning has also proven to be invaluable in pinpointing exactly where a tumour is and planning treatment with radiotherapy.
- The scanner is particularly good at testing for bleeding in the brain, for aneurysms, brain tumours and brain damage. It can also find tumours and abscesses throughout the body and is used to assess types of lung disease.
- In addition, the CT scanner is used to look at internal injuries such as a torn kidney, spleen or liver or bony injury, particularly in the spine. CT scanning can also be used to guide biopsies and therapeutic pain relieving procedures.

4.4 The Nature of Biomedical Signals

A signal is a function of one or several variables that carries useful information. A signal is said to be biological, if it is recorded from a living system and conveys information about the state or behaviour of that system. For example, the temperature record of a patient, the voltage record by an electrode placed on the scalp and the spatial pattern of X-ray absorption obtained from a CT scan are biological signals.

Origin of Biomedical Signals

Human body is made up of a number of systems, example, respiratory, cardiovascular, nervous system, etc. Each of these systems is made up of several sub-systems that carry on many physiological processes. Each physiological process is associated with certain types of signals referred as biomedical signals that reflect their nature and activities.

Different types of biomedical signals are,

- Biochemical signals, example: hormones and neurotransmitters.
- Bioelectrical signals, example: potentials and currents.
- Biomechanical signals, example: pressure and temperature.

Bioelectric signals are specific types of biomedical signals which are obtained by electrodes that record the variations in electrical potential generated by physiological processes.

Examples of bioelectric signals are,

- Electrocardiogram (ECG).
- Electroencephalogram (EEG).
- Electromyogram (EMG).
- Electrooculogram (EOG).

Observing these signals and comparing them to their known norms, any disease or disorder is detected. When such measurements are observed over a period of time, a one dimensional time series is obtained which is called a physiological signal. For example, when a person is suffering from a heart problem, it means that there are some changes in his electrocardiogram or changes in blood pressure. Another example is a neurological disorder, such as epilepsy, where there are changes in the patient's electroencephalogram.

Commonly used Biomedical Signals

The signals which are commonly used are:

- The electrocardiogram (ECG): It is the electrical activity of the heart or cardiac cells.

- The electroencephalogram (EEG): It is the electrical activity of the brain.
- The electromyogram (EMG): It is the electrical activity of the muscle cells.
- The electrogastogram (EGG): It is the electrical activity of the stomach.
- The phonocardiogram (PCG): It is the audio recording of the heart's mechanical activity.
- The carotid pulse (CP): It is the pressure of the carotid artery.
- The electoretinogram (ERG): It is the electrical activity of the retinal cells.
- The electrooculogram (EOG): It is the electrical activity of the eye muscles.

The bio-signals of electric origin are made up from integration of many action potentials. The action potential itself is the electric potential which is generated by a single cell, when it is mechanically, electrically or chemically stimulated.

4.4.1 Analog Signal Processing of Bio Signals: Amplifiers

A bio-amplifier is a device used to gather and increase the signal integrity of human neurophysiologic electrical activity for output to various sources.

Basic Requirement for Biological Amplifiers

- Bio-amplifiers must have high input impedance.
- Constant gain should be maintained throughout the required bandwidth.
- CMRR of bio-amplifiers should be more than 80 db.
- Drift free amplifiers can acts as good bio-amplifiers.
- It must have isolation and protection circuits.
- Voltage gain of bio-amplifiers should be more than 100 db.

Types of Bio-amplifiers

- Differential amplifier.
- Operational amplifier.
- Instrumentation amplifier.
- Chopper amplifier.
- Isolation amplifier.

Operational Amplifier

Operational amplifiers are highly stable, high gain DC difference amplifiers. Since there is no capacitive coupling between their various amplifying stages, they can handle signals from zero frequency up to a few hundred kHz. Their name is derived from the fact that they are used for performing mathematical operations on input signals.

There are two inputs, the inverting input (-) and the non-inverting input (+). These symbols have nothing to do with the polarity of the applied input signals.

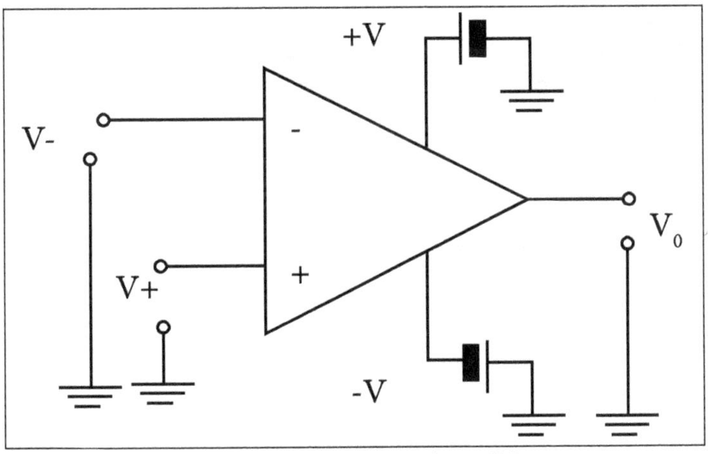

Symbol of the operational amplifier.

Inverting Amplifiers

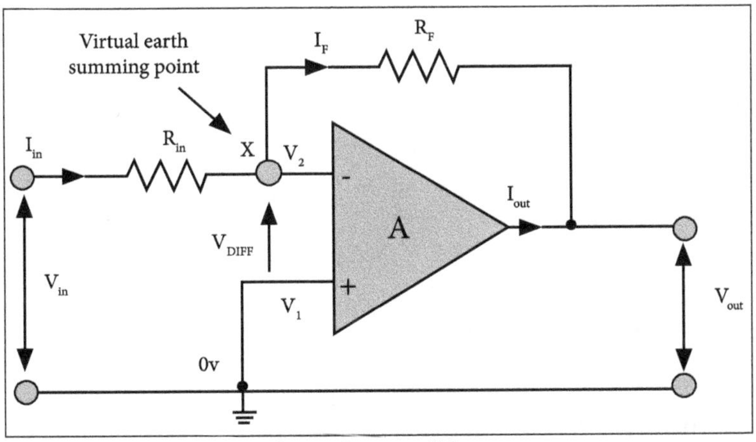

Inverting amplifier.

In inverting amplifier circuit, operational amplifier is connected with the feedback to produce a closed loop operation. While dealing with operational amplifiers, two important rules about inverting amplifiers should to considered. They are,

- No current flows into the input terminal.

- V_1 should always equal V_2 (i.e. $V_1 = V_2 = 0$).

However, in practical op-amp circuits, both of these rules are slightly broken.

This is because, the junction (x) of the input and feedback signal is at the same potential as the positive input, which is at zero volts or ground, then this junction is termed as virtual earth.

Due of this virtual earth node, input resistance of the amplifier is equal to the value of the input resistor, R_{in} and closed loop gain of the inverting amplifier can be set by the ratio of the two external resistors.

Then by using these two rules, we can derive the equation for calculating the closed loop gain of an inverting amplifier using the first principle.

Current (i) flows through the resistor network as shown below,

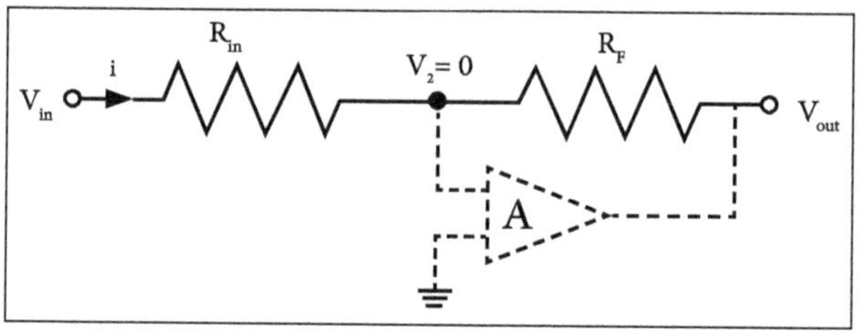

$$i = \frac{V_{in} - V_{out}}{R_{in} + R_f}$$

Therefore, $i = \frac{V_{in} - V_2}{R_{in}} = \frac{V_2 - V_{out}}{R_f}$

$$i = \frac{V_{in}}{R_{in}} - \frac{V_2}{R_{in}} = \frac{V_2}{R_f} - \frac{V_{out}}{R_f}$$

So, $\frac{V_{in}}{R_{in}} = V_2\left[\frac{1}{R_{in}} + \frac{1}{R_f}\right] - \frac{V_{out}}{R_f}$

And as, $i = \frac{V_{in} - 0}{R_{in}} = \frac{0 - V_{out}}{R_f} \quad \frac{R_f}{R_{in}} = \frac{0 - V_{out}}{V_{in} - 0}$

the Closed Loop Gain (Av) is given as $\frac{V_{out}}{V_{in}} = -\frac{R_f}{R_{in}}$

Then, the closed loop voltage gain of an inverting amplifier is given as,

$$\text{Gain}(Av) = \frac{V_{out}}{V_{in}} = -\frac{R_f}{R_{in}}$$

And this can be transposed to give V_{out} as,

$$V_{out} = -\frac{R_f}{R_{in}} \times V_{in}$$

The negative sign indicates that the output signal is inverted with respect to the input, as it is 180° out of phase. This is due to the feedback being negative in value.

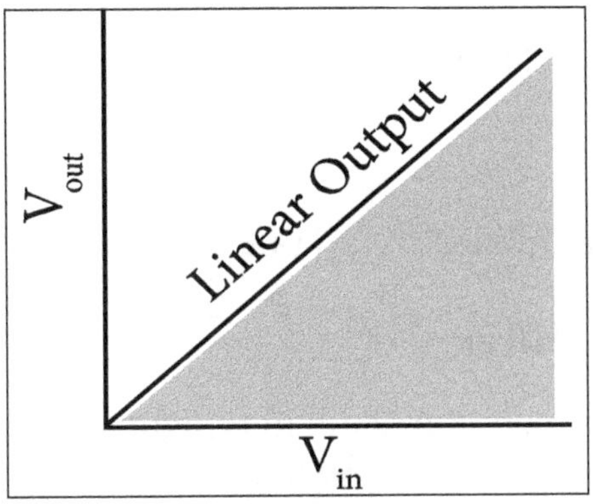

Linear output.

The equation for output voltage, V_{out} also shows that the circuit is linear in nature for a fixed amplifier gain as, $V_{out} = V_{in} \times \text{Gain}$.

This property can be used for converting a smaller sensor signal to a much larger voltage. It is also used in trans-resistance amplifier circuit.

Non-inverting Amplifiers

Non-inverting amplifier is one in which the output is in phase with respect to the input. Output is an non-inverted amplified version of the input.

Circuit Operation

The inverting amplifier using op-amp is shown in the below figure:

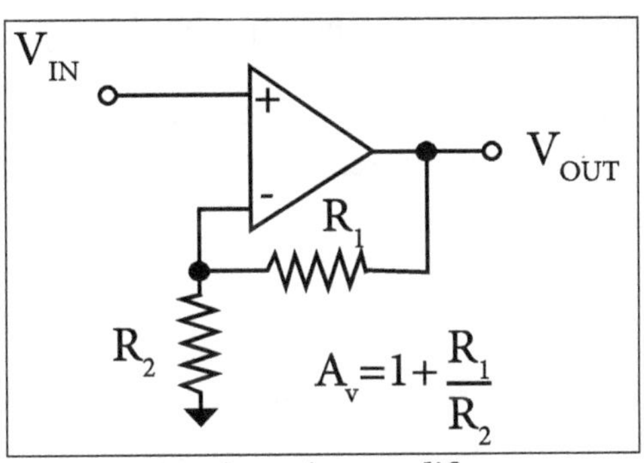

Non-inverting amplifier.

Assuming that the op-amp is ideal and applying the concept of virtual short, the voltage at the inverting terminal is equal to the non-inverting terminal.

Applying KCL at inverting node we get,

$$(V_{in} - V_{out})/R_2 + (V_{out} - 0)/R_1 = 0$$

By re-arranging the terms we get,

Voltage gain, $A_v = V_{out} / V_{in} = (1 + R_f / R_{in})$

Gain

Gain of the non-inverting amplifier, $A_v = (1 + R_f / R_{in})$

Instrumentation Amplifier

Physical quantities such as temperature, humidity, light intensity, water flow etc., are measured with the help of transducers. An instrumentation amplifier is used to amplify the output signal produced by a transducer, so that it can drive the indicator or display system.

Figure below shows the simplified form of such a system.

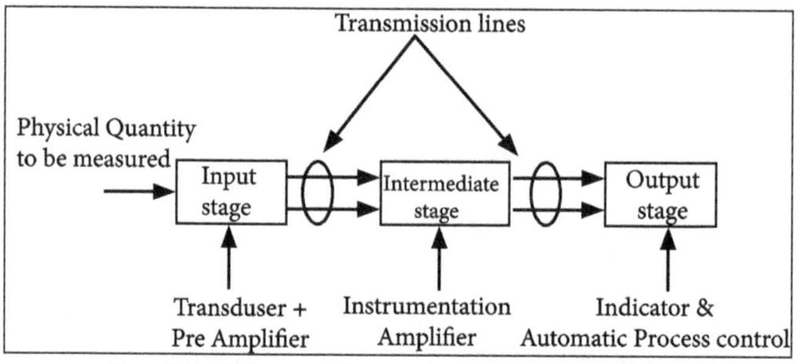

Simplified form of instrumentation amplifier.

The input stage is composed of a pre-amplifier and some sort of transducer, depending on the physical quantity to be measured. The output stage may use devices such as meter, oscilloscope, charts or magnetic recorders. Some transducer produce outputs with sufficient strength to permit their use directly, many do not.

Thus, instrumentation amplifiers are used in monitoring and control of physical quantities in industrial process for measurement and control of temperature, humidity and light intensity.

Requirements

Hence, a good instrumentation amplifier must meet the following specifications:

- High CMRR.

- High gain accuracy.
- High gain stability with low temperature co-efficient.
- Low AC offset.
- Low output impedance.

Specially designed op-amps such as μA725 are available to meet the above stated requirements of good instrumentation amplifier. Monolithic instrumentation amplifiers such as AD521, AD524 and AD624 are also commercially available.

Consider a basic differential amplifier as shown in the below figure:

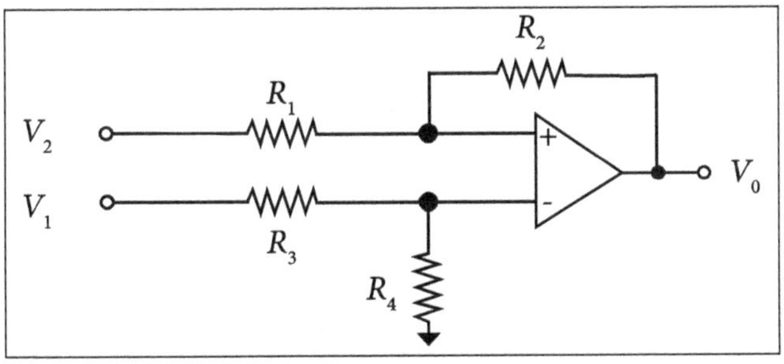

Differential amplifier using single op-amp.

It can be easily seen that the output voltage V_o is given by,

$$V_o = -\frac{R_2}{R_1} V_2 + \frac{1}{1+\left(\frac{R_3}{R_4}\right)} V_1 \left(1+\frac{R_2}{R_1}\right) \qquad ...(1)$$

Therefore,

$$V_o = -\frac{R_2}{R_1}\left[V_2 - \frac{1}{1+\left(\frac{R_3}{R_4}\right)}\left(1+\frac{R_1}{R_2}\right)V_1\right] \qquad ...(2)$$

If $\frac{R_1}{R_2} = \frac{R_3}{R_4}$, then $V_o = \frac{R_2}{R_1}(V_1 - V_2)$, (i.e.) the output voltage V_o is the difference of the two input voltages with the gain of R_2/R_1.

In the above circuit, V_1 has an input impedance $=(R_3 + R_4)$ and the impedance of source V_2 is only R_1.

This low impedance may load the signal source heavily. This limitation is overcome by the use of voltage follower between each input terminal and the difference amplifier.

An improved instrumentation amplifier results as shown in the below figure.

$$V_2' = R'I + V_2$$

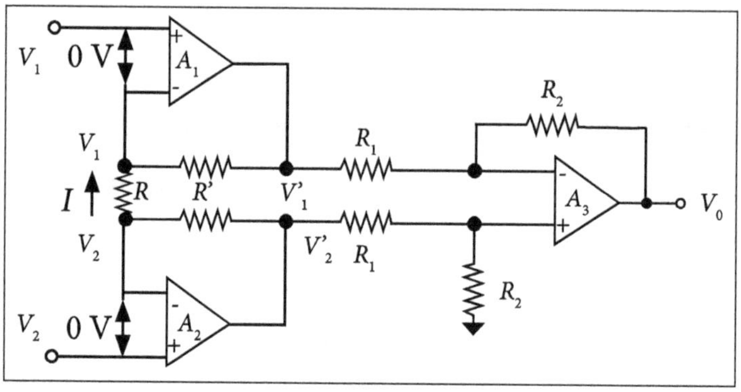

Instrumentation amplifier.

The op-amp A_1 and A_2 shown in the above figure are either voltage follower or buffer circuit, which act as the input stage for each of the inputs V_1 and V_2. The op-amps A_1 and A_2 have differential input voltages at zero.

Case 1: For $V_1 = V_2$, (i.e.) under common mode condition, the voltage across R will be zero.

The voltages at the inverting terminals of the buffers are equal to the input voltages. Since no current flows through R and R', the non-inverting amplifier acts as the voltage follower.

So its output $V_2' = V_2$. Similarly, op-amp A_2 acts as the voltage follower and has a output of $V_1' = V_1$.

Case 2: $V_1 \neq V_2$ current flows in R and R' and $\left[(V_2 - V_1') > (V_2 - V_1)\right]$.

Therefore, this circuit has more differential gain and CMRR when compared to the single op-amp circuit.

The output voltage V_o can be calculated as follows:

The voltage at the (+) input terminal of op-amp A_3 is $\dfrac{R_2 V_1'}{R_1 + R_2}$.

Using superposition theorem, we have,

$$V_o = -\frac{R_2}{R_1} V_1' + \left(1 + \frac{R_2}{R_1}\right)\left(\frac{R_2 V_2'}{R_1 + R_2}\right) \quad \text{...(3)}$$

On simplifying we get,

$$V_o = \frac{R_2}{R_1}(V_1' - V_2') \qquad ...(4)$$

Since, no current flows into the op-amp, the current flowing in R is $I = \frac{(V_1 - V_2)}{R}$, which passes through the resistor R'.

$$V_1' = R'I + V_1 = \frac{R'}{R}(V_1 - V_2) + V_1$$

$$V_2' = R'I + V_2 = -\frac{R'}{R}(V_1 - V_2) + V_2 \qquad ...(5)$$

Substituting the values of V_1' and V_2' in equation (4), we get,

$$V_o = \frac{R_2}{R_1}\left[\frac{2R'}{R}(V_2 - V_1) + (V_2 - V_1)\right]$$

$$V_o = \frac{R_2}{R_1}\left[1 + \frac{2R'}{R}\right](V_2 - V_1) \qquad ...(6)$$

By using a variable resistor R, the gain of this instrumentation amplifier can be varied. Since most of the instrumentation system uses the transducer in a bridge circuit, we will consider a simplified differential instrumentation system arrangement using a transducer bridge circuit.

Instrumentation Amplifier using Transducer Bridge

Figure below shows a simplified differential amplifier using transducer bridge. The circuit uses a resistance transducer whose resistance changes as a function of the physical quantity to be measured and is denoted by $(R_T \pm \Delta R)$,

Where,

R_T – Resistance of the transducer.

ΔR – Change in resistance.

The bridge is initially balanced by using DC supply voltage V_{dc}, so that $V_1 = V_2$.

(i.e.), $$\frac{R_B(V_{dc})}{R_B + R_A} = \frac{R_C(V_{dc})}{R_C + R_T}$$

Therefore,

$$\frac{R_A}{R_B} = \frac{R_T}{R_C}$$

Or,

$$R_T = \left(\frac{R_A}{R_B}\right)R_C$$

If the ratio of R_A to R_B, which is called the ratio arms of the bridge is assumed as constant k, then the value of the transducer resistance $R_T = kR_C$. As the physical quantity changes, the resistance R_T of the transducer also changes, causing an imbalance in the bridge. That is, output of the bridge $V_1 \neq V_2$.

As the resistor R_B and R_C are fixed resistors, the voltage V_1 is constant. However, the voltage V_2 varies as a function of the change in transducer resistance.

Therefore,

$$V_2 = \frac{R_C(V_{dc})}{R_C + (R_T + \Delta R)}$$

$$V_1 = \frac{R_B(V_{dc})}{R_C + R_B}$$

Consequently, the voltage V_{12} across the output terminal of the bridge is $V_{12} = V_2 - V_1$.

Therefore,

$$V_{12} = \frac{R_C(V_{dc})}{R_C + R_T + \Delta R} - \frac{R_B(V_{dc})}{R_C + R_B}$$

If $R_A = R_B = R_C = R_T = R$, then the above equation becomes,

$$V_{12} = \frac{\Delta R(V_{dc})}{2(2R + \Delta R)}$$

The gain of the basic differential amplifier is $-R_2/R_1$. Therefore, the output voltage V_o is given by,

$$V_o = V_{12}\left(-\frac{R_2}{R_1}\right) = \frac{\Delta R(V_{dc})}{2(2R + \Delta R)}\frac{R_2}{R_1}$$

Since the change in resistance of a transducer is normally very small, $(2R + \Delta R) \cong 2R$. Hence, the output voltage V_o becomes,

$$V_o = \left(\frac{R_2}{R_1}\right)\frac{\Delta R(V_{dc})}{4R}$$

Therefore, the output voltage is a function of the change in resistance of the transducer element, multiplied by the gain value of the op-amp A_3 and it is also determined by the resistor R.

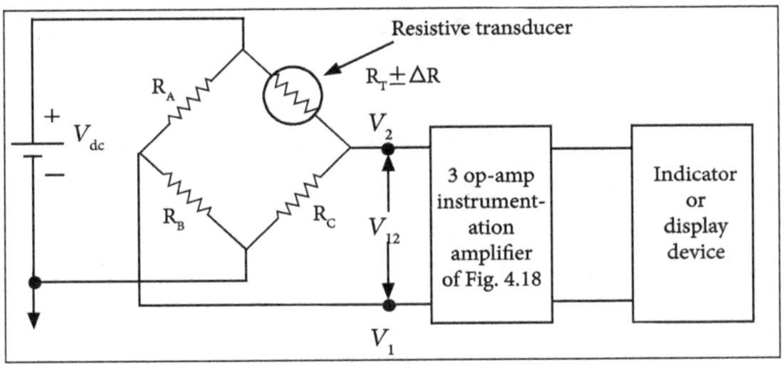

Instrumentation amplifier using transducer bridge.

Chopper Amplifiers

Analog signal is sampled in this amplifier circuit. Hence this circuit is known as chopper amplifier. Chopper amplifiers are classified into mechanical and non-mechanical choppers.

Advantages

- Provides low noise operation.
- Stable gain.

Isolation Amplifier

Isolation amplifier provides electrical isolation and an electrical safety barrier. They protect data acquisition components from common mode voltages, which are potential differences between instrument ground and signal ground. Amplifier with an isolation barrier allows the front-end of the amplifier to float with respect to common mode voltage to limit the barrier's breakdown voltage which is often 1,000 V_{dc}, peak AC or more.

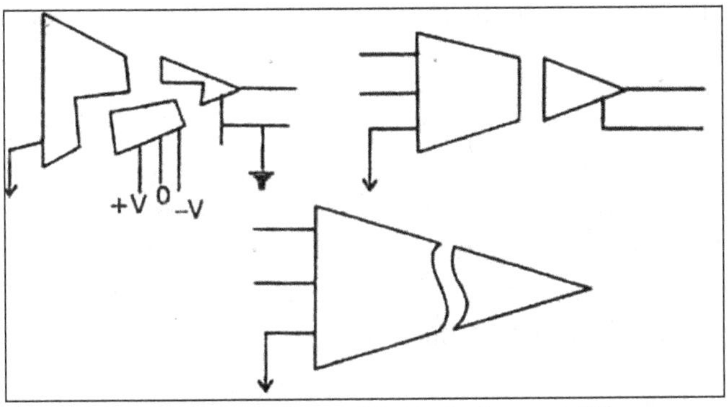

Different symbols used for isolation amplifier.

The isolation in the isolation amplifier is achieved by use of transformers or by use of optically coupled devices. Figure shown are the some of the symbols used for isolation amplifiers.

4.4.2 Transient Protection and Interference Reduction

The most critical point in the measurement of bio-potentials is the contact between electrodes and biological tissue. Both the electrode offset potential and the electrode/tissue impedance are subject to changes due to relative movements of electrode and tissue. Thus, two interference signals are generated as motion artifacts, (i.e.) the changes of the electrode potential and motion-induced changes of the voltage drop caused by the input current of the pre-amplifier.

These motion artifacts can be minimized by providing high input impedances for the preamplifier, usage of non-polarized electrodes with low half-cell potentials such as Ag/AgCl electrodes and by reducing the source impedance by the use of electrode gel.

Motion artifacts, interferences from external electromagnetic fields and noise can also be generated in the wires connecting electrodes and amplifier. Reduction of these interferences is achieved by using twisted pair cables, shielded wires and input guarding.

Recording of bio-potentials is often done in an environment that is equipped with many electrical systems which produce strong electrical and magnetic fields. In addition to 60 Hz power line frequency and some strong harmonics, high frequency electromagnetic fields are encountered. At power line frequency, the electric and magnetic components of the interfering fields can be considered separately. Electrical fields are caused by all conductors that are connected to power, even with no flow of current.

A current is capacitively coupled into the body, where it flows to the ground electrode. If an isolation amplifier is used without patient ground, the current is capacitively coupled to the ground. In this case, the body potential floats with a voltage of up to 100 V towards the ground. Minimizing interferences requires increasing the distance between power lines and the body, use of isolation amplifiers, separate grounding of the body at a location as far away from the measuring electrodes as possible and use of shielded electrode cables.

The magnetic field components produce eddy currents in the body. The amplifier, the electrode cable and the body form an induction loop that is subject to the generation of an interference signal. Minimizing this interference signal requires increasing the distance between the interference source and patient, twisting the connecting cables, shielding of the magnetic fields and relocating the patient to a place and orientation that offers minimum interference signals.

In many cases, an additional narrow band rejection filter or notch filter is implemented as an additional stage in the bio-potential amplifier to provide sufficient suppression of line frequency interferences.

In order to achieve optimum signal quality, the bio-potential amplifier has to be adapted to the specific application. Based on the signal parameters, both appropriate bandwidth and gain factors are chosen.

The below figure shows an overview of the most commonly measured bio-potentials and specifies the normal ranges for amplitude and bandwidth.

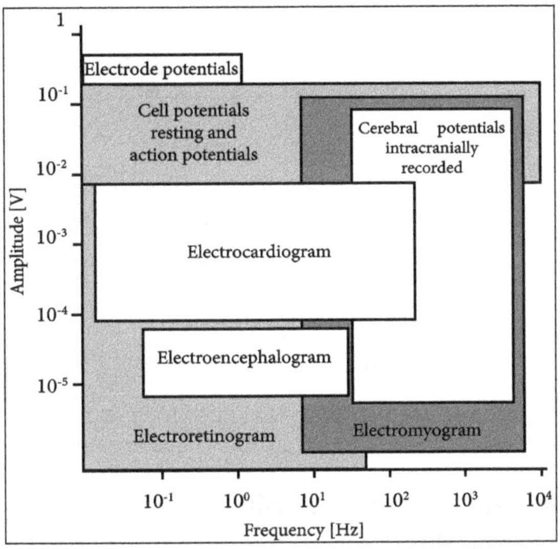

Amplitudes and spectral ranges of some important bio-signals. The various bio-potentials completely cover the area from 10-6 V to almost 1 V and from DC to 10 kHz.

A final requirement for bio-potential amplifiers is the need for calibration. Since the amplitude of the bio-potential often has to be determined very accurately, there must be provisions to easily determine the gain or the amplitude range referenced to the input of the amplifier. For this purpose, the gain of the amplifier must be well calibrated.

In order to prevent difficulties with calibrations, some amplifiers that need to have adjustable gain use a number of fixed gain settings rather than providing a continuous gain control. Some amplifiers have a standard signal source of known amplitude built in that can be momentarily connected to the input by the push of a button to check the calibration at the output of the bio-potential amplifier.

4.4.3 Movement Artifact Circuits

Often a signal is disturbed by artifacts and most of these are called the movement artifacts. Movement artifacts originate from two different phenomena:

- Movement of the cables.
- Movement of the electrode.

The movement artifacts from the cables can be solved in a very simple way, (i.e.) by shielding the cables with the electrode signal itself.

Movement artifacts from the electrode are often measured. They are caused by the pressure, which is exerted on the gel layer of the electrode. That pressure modifies the double layer between the tissue and the metal of the electrode itself. This double layer is responsible for the offset of the electrode, so the offset of the electrode is modified by any pressure that is exerted on the surface of the interface.

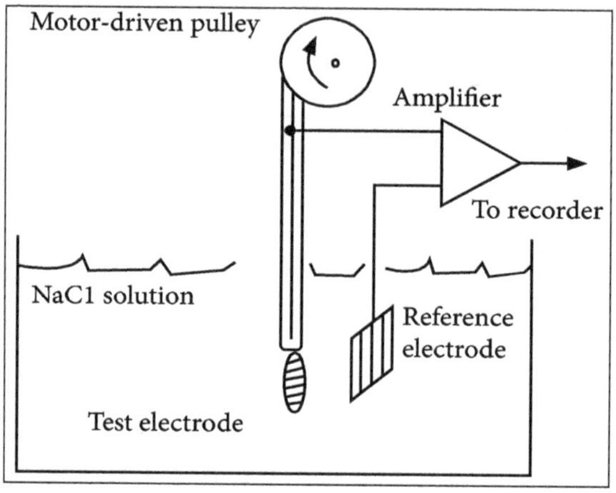

Test setup for motion artifact measurements.

The amount of movement artifact can be minimized by using a special type of electrode, the so called microelectrodes. When an electrode is used with very small amount of gel and a very small surface for the measurement, the movement artifact is reduced considerably. The microelectrodes have much higher impedance than the standard ECG electrodes, so using this type of electrode a special amplifier system with very high input impedance is required.

4.5 Active filters and Rate Measurement

Active First Order Low Pass Filter

The most common active filter is the active low pass filter. Its principle of operation and frequency response is exactly the same as those of passive filter, the only difference is that it uses an op-amp for gain control and amplification. The simplest form of a low pass active filter is to connect an inverting or non-inverting amplifier.

This first order active low pass filter, consists of a passive RC filter stage providing a low frequency path to the input of a non-inverting operational amplifier. The amplifier is configured as a voltage follower giving it a DC gain of unity, (i.e.) $A_v = +1$ as opposed to the passive RC filter, which has a DC gain of less than unity.

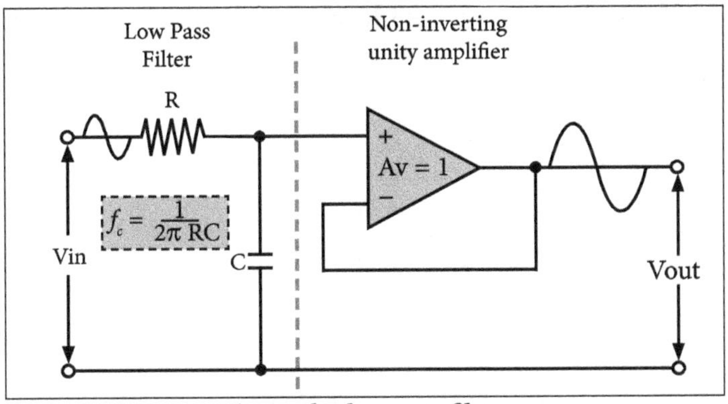

First order low pass filter.

The advantage of this configuration is that the op-amps high input impedance prevents excessive loading on the filters output, while its low output impedance prevents the filters cut-off frequency point from being affected by the changes in the impedance of the load.

This configuration provides good stability to filters, but its main disadvantage is that it has no voltage gain above one. However, although the voltage gain is one, the power gain is very high as its output impedance is much lower than its input impedance. Whenever, a voltage gain greater than one is required we can use the following filter circuit.

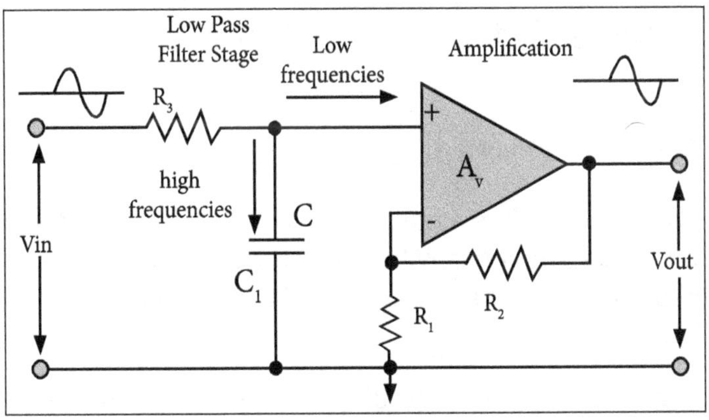

Active low pass filter with amplification.

The frequency response of the circuit will be the same as that for the passive RC filter, except that the amplitude of the output is increased by the pass band gain, A_F of the amplifier. For a non-inverting amplifier circuit, the magnitude of voltage gain for the filter is given as a function of the feedback resistor (R_2) divided by its corresponding input resistor (R_1).

This can be written as,

$$\text{DC gain} = \left(1 + \frac{R_2}{R_1}\right)$$

Therefore, the gain of an active low pass filter as a function of frequency will be,

Gain of a first order low pass filter,

$$\text{Voltage Gain,}(A_v)=\frac{V_{out}}{V_{in}}=\frac{A_F}{\sqrt{1+\left(\frac{f}{f_c}\right)^2}}$$

Where,

A_F = Pass band gain of the filter, $(1+R_2/R_1)$.

f = Frequency of the input signal in Hertz, (Hz).

f_c = Cut-off frequency in Hertz, (Hz).

Thus, the operation of a low pass active filter can be verified from the frequency gain equation.

- At very low frequencies, $f < f_c$ $\frac{V_{out}}{V_{in}} \cong A_F$.
- At cut-off frequency, $f = f_c$ $\frac{V_{out}}{V_{in}} = \frac{A_F}{\sqrt{2}} = 0.707\ A_F$.
- At very high frequencies, $f > f_c$ $\frac{V_{out}}{V_{in}} < A_F$.

Thus, the active low pass filter has a constant gain A_F from 0 Hz to the high frequency cut-off point, f_c. At f_c, the gain is $0.707A_F$ and after f_c it decreases at a constant rate as the frequency increases, (i.e.) when frequency is increased by ten times, the voltage gain is divided by 10.

In other words, the gain decreases by 20 dB each time when the frequency is increased by 10. When dealing with filter circuits, the magnitude of the pass band gain of the circuit is generally expressed in decibels or dB as a function of the voltage gain and this is defined as,

Magnitude of voltage gain in (dB),

$$A_V(dB)=20\log_{10}\left(\frac{V_{out}}{V_{in}}\right)$$

$$\therefore -3dB=20\log_{10}\left(0.707\frac{V_{out}}{V_{in}}\right)$$

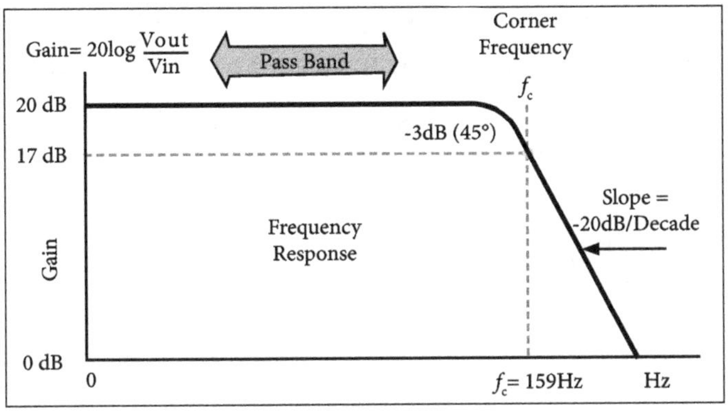

Frequency response curve.

If the external impedance connected to the input of the circuit changes, this change will also affect the corner frequency of the filter. One way of avoiding this is to place the capacitor in parallel with the feedback resistor R_2.

The value of the capacitor will slightly change from 100 nF to 110 nF. The formula used to calculate the cut-off corner frequency is the same as that used for the RC passive low pass filter.

$$f_C = \frac{1}{2\pi CR_2} \text{Hertz}$$

Active Second Order Low Pass Filter

As with the passive filter, a first order low pass active filter can be converted into a second order low pass filter by simply using an additional RC network in the input path. The frequency response of the second order low pass filter is identical to that of the first order except that the stop band roll-off will be twice the first order filters at 40 dB/decade. Hence, the design steps required for second order active low pass filter are the same.

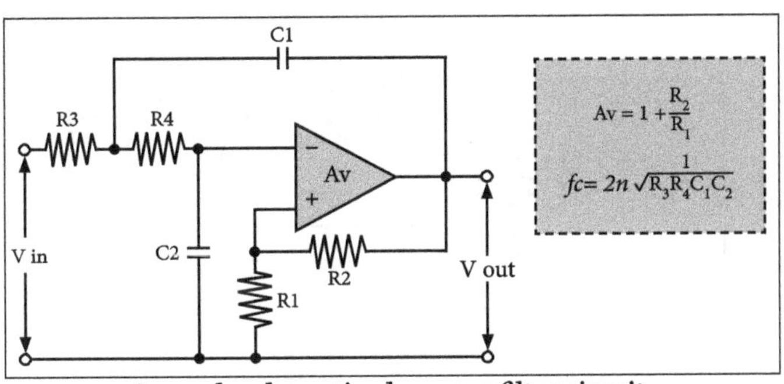

Second order active low pass filter circuit.

When cascading filter circuits together to form higher order filters, the overall gain of the filter is equal to the product of each stage. For instance, the gain of the first stage may be 10 and the gain of the second stage may be 32 and the gain of the third stage may be 100. Then, the overall gain will be 32,000, as shown in the below figure:

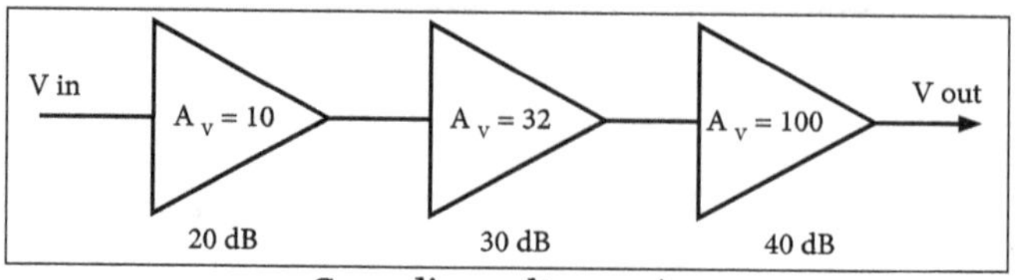

Cascading voltage gain.

$$A_v = A_{v1} \times A_{v2} \times A_{v3}$$

$$A_v = 10 \times 32 \times 100 = 32000$$

$$A_v(dB) = 20\log_{10}(32,\ 000)$$

$$A_v(dB) = 90\ dB$$

$$90\ dB = 20\ dB + 30\ dB + 40\ dB$$

Second order active filters are important, because higher order filters can be designed using them. By cascading together the first and second order filters, filters with an order value, either odd or even up to any value can be constructed.

Active First Order High Pass Filters

The basic electrical operation of an active high pass filter is exactly the same as its equivalent RC passive high pass filter circuit, except that the circuit has an operational amplifier included within its filter design providing gain and amplification control.

Like the active low pass filter circuit, the simplest form of an active high pass filter is to connect a standard inverting or non-inverting op-amp to the basic RC high pass passive filter circuit as shown below:

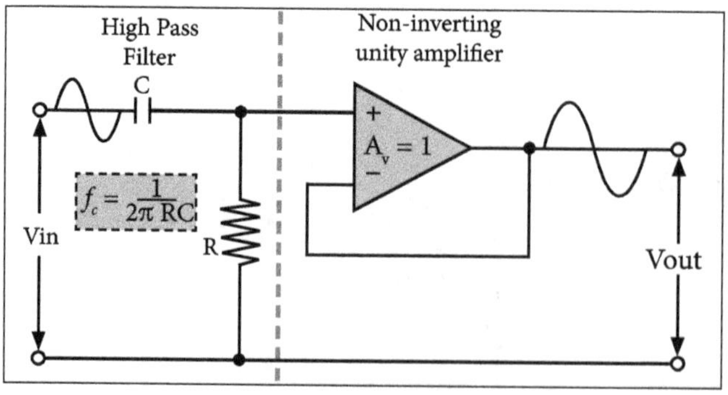

First order high pass filter.

Unlike passive high pass filters, which have an infinite frequency response, the maximum pass band frequency response of an active high pass filter is limited by the bandwidth or open loop characteristics of the operational amplifier being used, making them appear as if they are band pass filters with a high cut-off frequency, determined by the selection of op-amp and gain.

The maximum frequency response of an op-amp is limited to the gain/bandwidth product or open loop voltage gain (A_V) of the op-amp being used giving it a bandwidth limitation, where the closed loop response of the op-amp intersects the open loop response.

A commonly available operational amplifier such as the μA741 has a typical open loop DC voltage gain of about 100 dB reducing at a roll-off rate of -20 dB/decade as the input frequency increases. The gain of the μA741 reduces until it reaches unity gain or its transition frequency (f_t) which is about 1 MHz. This causes the op-amp to have a frequency response curve very similar to that of a first order low pass filter and this is shown below.

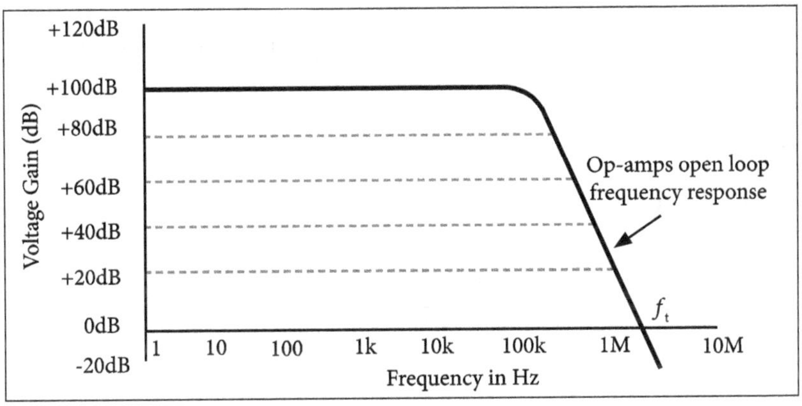

Frequency response curve of a typical op-amp.

Then, the performance of a high pass filter at high frequencies is limited by this unity gain cross-over frequency, which determines the overall bandwidth of the open loop amplifier. The gain bandwidth product of the op-amp starts from around 100 kHz for small signal amplifiers up to about 1 GHz for high speed digital video amplifiers. op-amp based active filters can achieve good performance and very good accuracy, provided that low tolerance resistors and capacitors are used.

Under normal circumstances, the maximum pass band required for a closed loop active band pass or high pass filter is well below that of the maximum open loop transition frequency. However, when designing active filter circuits it is important to choose the correct op-amp for the circuit, as the loss of high frequency signals may result in signal distortion.

A first order active high pass filter attenuates low frequencies and passes high frequency signals. It simply consists of a passive filter section followed by a non-inverting operational amplifier. The frequency response of the circuit is the same as that of the passive filter, except that the amplitude of the signal is increased by the gain of the amplifier and for a non-inverting amplifier the value of the pass band voltage gain is given as, 1 + R_2/R_1, the same as that for the low pass filter circuit.

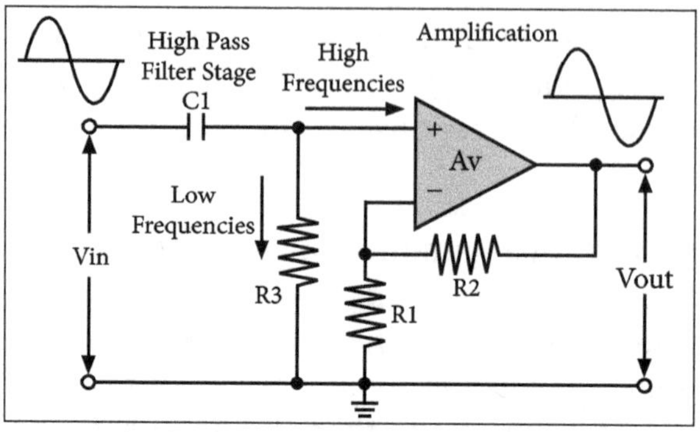

Active high pass filter with amplification.

For a non-inverting amplifier circuit, the magnitude of the voltage gain for the filter is given as a function of the feedback resistor (R_2) divided by its corresponding input resistor (R_1).

This can be written as,

Gain for an active high pass filter,

$$\text{Voltage Gain,}\left(A_V\right)=\frac{V_{out}}{V_{in}}=\frac{A_F\left(\frac{f}{f_c}\right)}{\sqrt{1+\left(\frac{f}{f_c}\right)^2}}$$

Where,

A_F = Pass band gain of the filter, $\left(1+R_2/R_1\right)$.

f = Frequency of the input signal in Hertz, (Hz).

f_c = Cut-off frequency in Hertz, (Hz).

Just like the low pass filter, the operation of a active high pass filter can be verified from the frequency gain equation.

- At very low frequencies, $f < f_c \frac{V_{out}}{V_{in}} < A_F$.
- At cut-off frequency, $f = f_c \frac{V_{out}}{V_{in}} = \frac{A_F}{\sqrt{2}} = 0.707\ A_F$.
- At very high frequencies, $f > f_c \frac{V_{out}}{V_{in}} \cong A_F$.

Then, the active high pass filter has a gain A_F that increases from 0 Hz to the low frequency cut-off point, f_c at 20 dB/decade, as the frequency increases. At f_c, the gain is $0.707A_F$ and after f_c, all frequencies are pass band frequencies, so the filter has a constant gain A_F, with the highest frequency being determined by the closed loop bandwidth of the op-amp.

Generally, when dealing with filter circuits, the magnitude of the pass band gain of the circuit is expressed in dB as a function of the voltage gain and this is defined as,

Magnitude of voltage gain in (dB),

$$A_V(dB) = 20 \log_{10}\left(\frac{V_{out}}{V_{in}}\right)$$

$$\therefore -3dB = 20 \log_{10}\left(0.707 \frac{V_{out}}{V_{in}}\right)$$

For a first order filter, the frequency response curve of the filter increases by 20 dB/ decade or up to the determined cut-off frequency point, which is always at -3 dB below the maximum gain value. The lower cut-off or corner frequency (fc) can be found by using the below formula:

$$f_c = \frac{1}{2\pi RC} Hz$$

The corresponding phase shift or phase angle of the output signal is the same as that given for the passive RC filter. It is equal to +45° at the cut-off frequency f_c and is given by,

$$\text{Phase shift } \phi = \tan^{-1}\left(\frac{1}{2\pi f RC}\right)$$

A simple first order active high pass filter can also be made by using an inverting operational amplifier configuration.

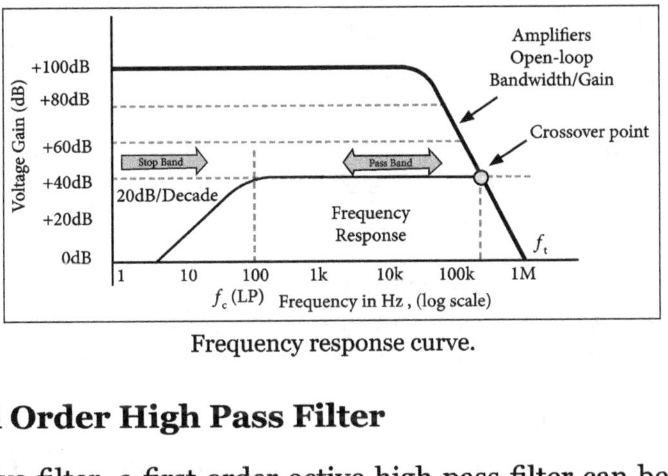

Frequency response curve.

Active Second Order High Pass Filter

As with the passive filter, a first order active high pass filter can be converted into a second order high pass filter by simply using an additional RC network in the input path. The frequency response of the second order high pass filter is identical to that of the first order, except that the stop band roll-off will be twice the first order filter at 40 dB/decade. Therefore, the design steps required for the active second order high pass filter are the same.

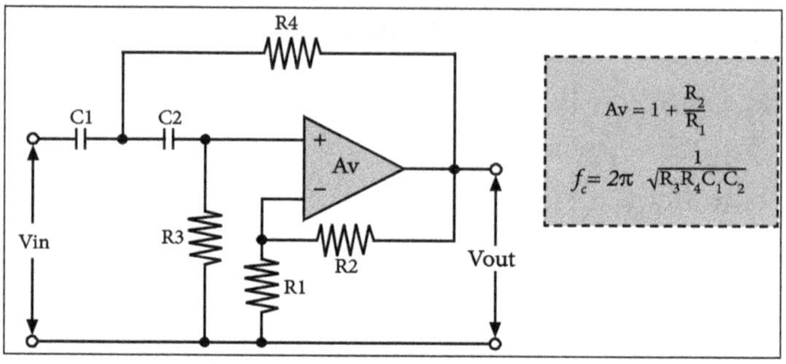

Second order active high pass filter circuit.

Higher order active high pass filters, such as third, fourth, fifth, etc., are formed by simply cascading together the first and second order filters. For example, a third order high pass filter is formed by cascading in series the first and second order filters, a fourth order high pass filter is obtained by cascading two second order filters together and so on.

Then, an active high pass filter with an even order number will consist of only second order filters, whereas an odd order number will start with a first order filter at the beginning as shown below:

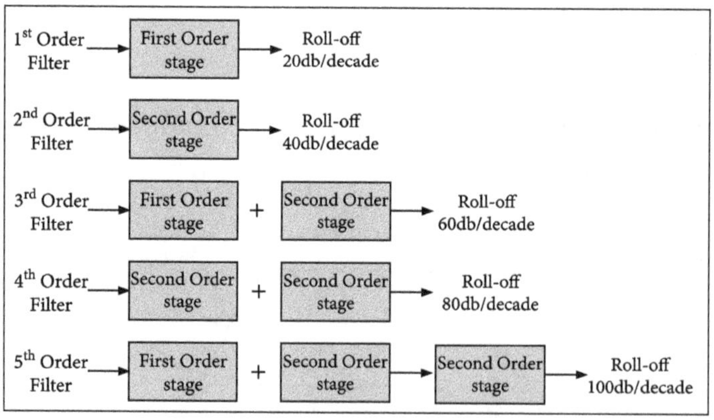

Cascading active high pass filters.

Although, there is no limit to the order of a filter that can be formed, as the order of the filter increases to its size. Also, its accuracy declines, that is the difference between the actual stop band response and theoretical stop band response increases.

If the frequency determining resistors are equal, $R_1 = R_2 = R_3$ etc. and the frequency determining capacitors are equal, $C_1 = C_2 = C_3$ etc., then the cut-off frequency for any order of the filter will exactly be the same. However, the overall gain of the higher order filter is fixed, because all the frequency determining components are equal.

4.5.1 Averaging and Integrator Circuits

Summing Amplifiers

The summing amplifier is a very flexible circuit. As its name suggests, the summing amplifier can be used for combining the voltage present on two or more inputs into a single output voltage.

If we add more resistors to the input, each equal in value to the original input resistor, R_{in} we end up with another operational amplifier circuit called a summing inverter, summing amplifier or even a voltage adder circuit as shown below:

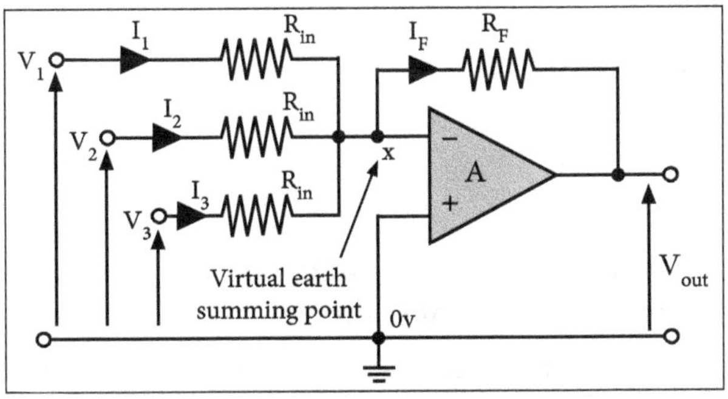

Summing inverter.

The output voltage (V_{out}) now becomes proportional to the sum of the input voltages, V_1, V_2, V_3 etc. Then, we can modify the original inverting amplifier equation by considering these new inputs. Thus,

$$I_F = I_1 + I_2 + I_3 = -\left[\frac{V_1}{R_{in}} + \frac{V_2}{R_{in}} + \frac{V_3}{R_{in}}\right]$$

$$\text{Inverting Equation}: V_{out} = -\frac{R_f}{R_{in}} \times V_{in}$$

$$\text{Then, } -V_{out} = \left[\frac{R_F}{R_{in}}V_1 + \frac{R_F}{R_{in}}V_2 + \frac{R_F}{R_{in}}V_3\right]$$

However, if all the input impedances (R_{in}) are equal in value, we can simplify the above equation to give the output voltage of the summing amplifier,

The equation of the summing amplifier is given as,

$$-V_{out} = \frac{R_F}{R_{IN}}(V_1 + V_2 + V_3 \text{....etc})$$

We now have an op-amp that will amplify each individual input voltage and produce an output voltage signal that is proportional to the algebraic sum of the three individual input voltages V_1, V_2 and V_3. We can also add more inputs if required, as each individual input sees their respective resistance, R_{in} as the only input impedance.

This is because the input signals are effectively isolated from each other by the virtual earth node at the inverting input terminal of the op-amp. A direct voltage addition can also be obtained when all the resistances are of equal value and R_f is equal to R_{in}.

A scaling summing amplifier can be made if the individual input resistors are not equal. Then, the equation will be modified to,

$$-V_{out} = V_1\left(\frac{R_f}{R_1}\right) + V_2\left(\frac{R_f}{R_2}\right) + V_3\left(\frac{R_f}{R_3}\right) \text{....etc}$$

We can re-arrange the above equation as,

$$-V_{out} = R_f\left(\frac{V_1}{R_1} + \frac{V_2}{R_2} + \frac{V_3}{R_3}\right) \text{....etc}$$

This allows the output voltage to be calculated easily if more input resistors are connected to the inverting input terminal of the amplifier. The input impedance of each individual channel is the value of their respective input resistors, (i.e.) R_1, R_2, R_3 ... etc.

Sometimes, we need a summing circuit to add two or more voltage signals together without any amplification. By putting all the resistances of the circuit above the same value R, the op-amp will have a voltage gain of unity and an output voltage equal to the direct sum of all the input voltages as shown in the below figure:

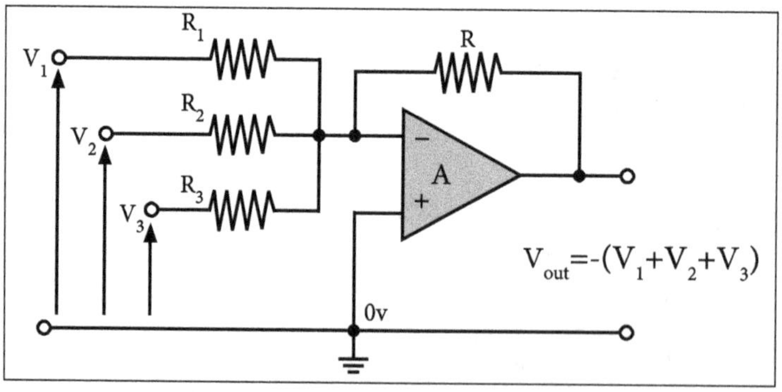

Op-amp with voltage gain of unity.

The summing amplifier is a very flexible circuit, enabling us to effectively add together several individual input signals. If the inputs resistors, R_1, R_2, R_3, etc., are equal, a unity gain inverting adder will be made. However, if the input resistors are of different values, a scaling summing amplifier is produced, which will output a weighted sum of the input signals.

Integrator

The process of integration is complementary to that of differentiation. A circuit in which the output voltage waveform is the integral of the input voltage waveform is the integrator. Such a circuit is obtained by using a basic inverting amplifier configuration with the feedback resistor R_f replaced by a capacitor C_f.

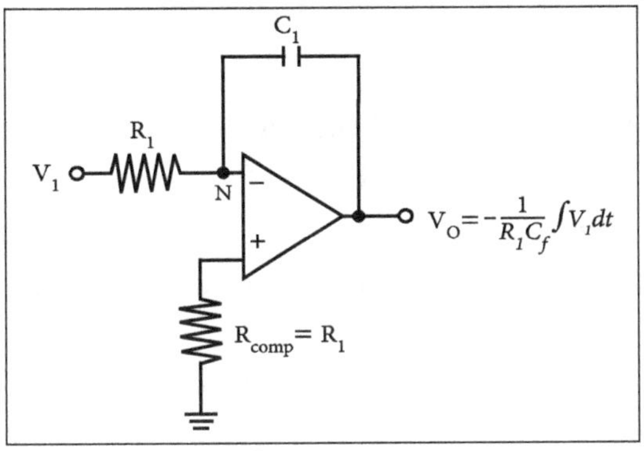

Integrator.

Analysis

The nodal equation at node N is,

$$\frac{v_i}{R_1} + C_f \cdot \frac{dv_o}{dt} = 0$$

$$C_f \frac{dv_o}{dt} = \frac{-v_i}{R_1}$$

$$\frac{dv_o}{dt} = -\frac{v_i}{C_f R_1}$$

Integrating both sides, we get

$$\int_o^t \frac{dv_o}{dt} = -\frac{1}{R_1 C_f} \int_o^t v_i dt$$

$$v_o(t) = \frac{1}{R_1 C_f} \int_o^t v_i(t) dt + v_o(o) \qquad ...(1)$$

Where $v_o(o)$ is the initial output voltage. Hence, the output is $1/(R_1 C_f)$ times the integral of input and $R_1 C_f$ is the time constant of the integrator.

The AC or Continuous Op-amp Integrator

If we change the above square wave input signal to that of a sine wave of varying frequency the Op-amp Integrator performs less like an integrator and begins to behave more like an active "Low Pass Filter", passing low frequency signals while attenuating the high frequencies.

At 0Hz or DC, the capacitor acts like an open circuit blocking any feedback voltage resulting in very little negative feedback from the output back to the input of the amplifier. Then with just the feedback capacitor, C, the amplifier effectively is connected as a normal open-loop amplifier which has very high open-loop gain resulting in the output voltage saturating.

This circuit connects a high value resistance in parallel with a continuously charging and discharging capacitor. The addition of this feedback resistor, R_2 across the capacitor, C gives the circuit the characteristics of an inverting amplifier with finite closed loop gain of R_2 / R_1. At very low frequencies, the circuit acts as an standard integrator, whereas at higher frequencies, the capacitor shorts out the feedback resistor, R_2 due to the effects of capacitive reactance reducing the amplifier gain.

The AC Op-amp Integrator with DC Gain Control

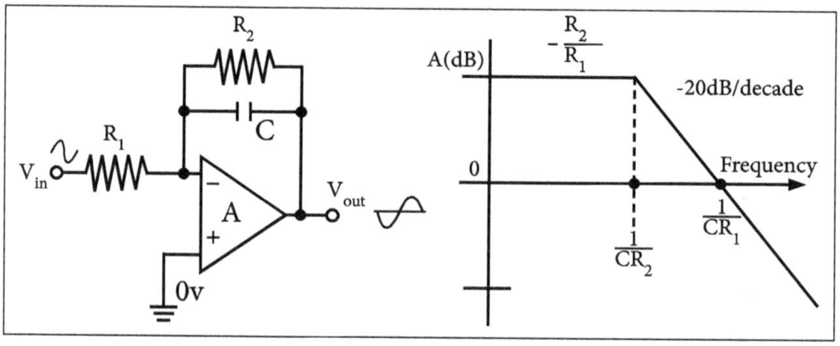

Unlike the DC integrator, amplifier above whose output voltage at any instant will be the integral of a waveform so that when the input is a square wave, the output waveform will be triangular. For an AC integrator, a sinusoidal input waveform will produce another sine wave as its output which will be 90° out of phase with the input producing a cosine wave.

Further, when the input is triangular, the output waveform is also sinusoidal. Then, this forms the basis of a active low pass filter as,

$$\text{D.C. Voltage Gain, } (Av_o) = -\frac{R_2}{R_1}$$

$$\text{A.C. Voltage Gain, } (Av) = -\frac{R_2}{R_1} \times \frac{1}{(1 - 2\pi f C R_2)}$$

$$\text{Corner Frequency, } (f_o) = \frac{1}{2\pi C R_2}$$

4.5.2 Transient Protection Circuits

Transients can damage or disrupt the products connected to the signal or power lines. Common sources of transient energy coupled from lightning, electrostatic discharge and circuits experience a sudden change in current due to opening of switch or short circuit.

Transient protection devices attempt to re-direct the energy in these transients by taking advantage of the differences between the transient waveform and the intended power or signal waveform. The most common transient protection schemes limit the voltage amplitude, current amplitude or transition times on the circuit they are protecting.

Voltage Limiting Devices

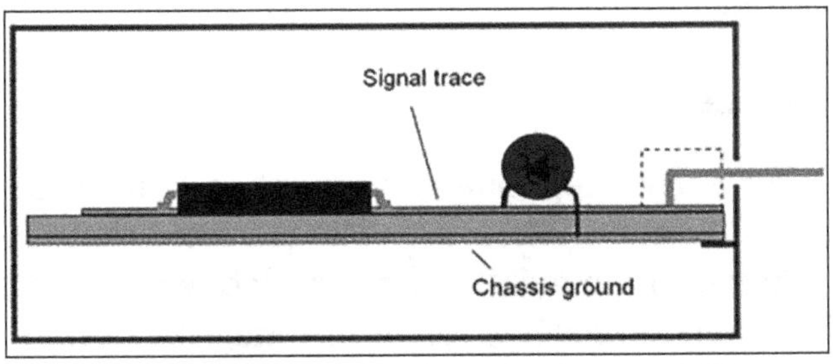

Voltage limiting transient protection on a PCB.

Figure illustrates how a voltage limiting transient protection device might be used to protect an input of a VLSI component mounted on a printed circuit board. At normal signal or power voltages, the device has high impedance and does not significantly affect the operation of the circuit. However, if the voltage across the device exceeds a threshold, the impedance across its terminals decreases suddenly, diverting current from the component being protected.

Diodes

Diodes are the most common protection device for low voltage applications. Generally, an unbiased diode has high impedance when the potential across its terminals is below about 0.5 V. The impedance drops quickly at higher voltages. Diodes can be combined in series to reach higher threshold voltages or reversed biased zener diodes can be employed. Generally, diodes are used in applications requiring a threshold voltage of anywhere from 0.5 to a few volts.

Diodes are voltage limiting devices, which means they conduct just enough current to hold the voltage to the threshold level. They are relatively fast acting with response times fast enough to protect sensitive semiconductor inputs. However, diodes are incapable of dissipating much energy before they fail. A failed diode may look like an open circuit or a short circuit, but is more likely to fail short.

Varistors

Varistors are another type of voltage limiting device, made from metal oxide powder. Varistors can be designed to have threshold voltages anywhere from 0.5 V to 10 V. They are capable of dissipating more energy than the diodes, but they are also likely to have more parasitic capacitance, which may make them unsuitable for high speed signal applications. Like diodes, they are more likely to fail short than open.

Thyristors

Thyristors are semiconductor devices like diodes. However, they are generally capable of dissipating much more energy than diodes and they can be found with a variety of threshold voltages. Unlike diodes and varistors, thyristors are a crowbar device, which means that they effectively short out when their threshold voltage is exceeded and the voltage across them drops to zero.

Gas Discharge Devices

Lightning protection for early telephone equipment in homes consist of two pointed piece of metal in close proximity. When the voltage across these metal pieces exceeds the threshold, the air between the metal pieces break down forming an arc. This is still an effective transient protection scheme for large voltages, but the devices that are used today are enclosed in a glass or plastic tube filled with a gas that breaks down more predictably than air.

Gas discharge tubes are capable of dissipating relatively large amounts of energy without being damaged. They also have relatively low capacitance, so they are less likely to distort fast. Generally, they are designed with threshold voltages anywhere from 10 to 100 V and are crowbar devices like thyristors.

Gas discharge tubes are likely to fail open making it difficult to tell whether they are working properly. However, a failed discharge tube will not interfere with normal operation of the device. Neon bulbs behave like gas discharge tubes and can be an inexpensive way to provide primary transient protection for many applications.

Current Limiting Devices

Devices such as fuses, circuit breakers and thermal protection devices are triggered by current rather than voltage. These devices are placed in series with the power or signal lines leading to the equipment being protected. Normally they have a very low impedance, but open when too much current flows through them.

Once they are triggered, current limiting devices block power from reaching the equipment being protected without having to dissipate that power as heat. Therefore, there is virtually no limit to the amount of energy or power that they can handle. However, current limiting devices do not generally respond fast enough to protect equipment from the fast transients generated by lightning or electrostatic discharge.

Transition Time Limiting Devices

Voltage and current limiting devices take a finite amount of time to respond. If the transient is fast, the damage may occur before the protection device has a chance to operate. The best transient protection is a simple capacitor or ferrite bead that is designed to slow any change in voltage or current due to an induced transient.

Figure below illustrates how a capacitor on a sensitive input to a VLSI component can slow the rise time associated with any transients induced. Integrated circuit inputs will respond to very fast transients even when this is not required for the proper operation of the device.

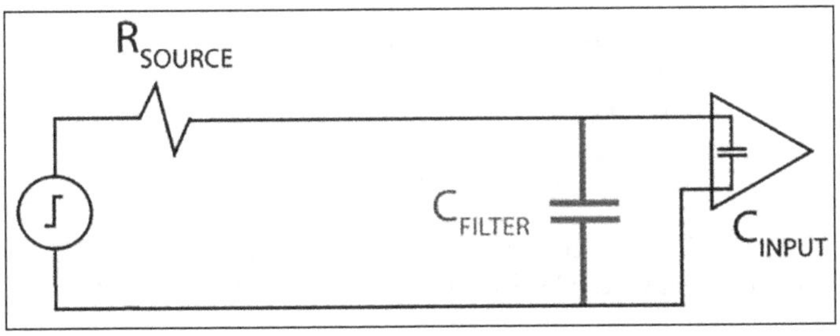

Using a filter capacitor to slow the response of a fast input.

For example, the reset input on a microprocessor is not generally switched on a frequent basis. When it is switched, it normally does not matter if the switching occurs in microseconds or milliseconds. These inputs often respond to transients on the order of nanoseconds. Slowing these inputs by adding a shunt capacitor, can eliminate problems due to very fast transients without having any adverse effect on the operation of the device.

$$\text{Rise time} = 2.2\, R_{SOURCE}\left(C_{FILTER} + C_{INPUT}\right)$$

Capacitors have several advantages over other transient protections devices. They are relatively small, inexpensive and their linear behaviour is relative easy to predict and model. They have a relatively large energy storage capacity compared to the devices they protect, so they are not likely to fail if properly sized. Although, capacitors generally behave like short circuits when exposed to voltages higher than their rated value, their behaviour under these conditions is not reliable, so capacitors should not be used as voltage limiting transient protection devices.

Ferrite beads or resistors can be used to provide transition time limiting for low impedance inputs. Ferrite beads have the advantage that, no DC voltage is dropped across them. However, when using a ferrite bead, it is important to ensure that the signal or power currents do not saturate the ferrite material.

4.6 Time-frequency Representations

Time Frequency Representations (TFR) combines the time domain and frequency domain representations into a single framework to obtain the notion of time frequency. TFR offer the time localization vs. frequency localization trade-off between two extreme cases of time domain and frequency domain representations. The Short Time Fourier Transform (STFT) and the Gabor transform are the classical examples of linear time frequency transforms which use time shifted and frequency shifted basis functions.

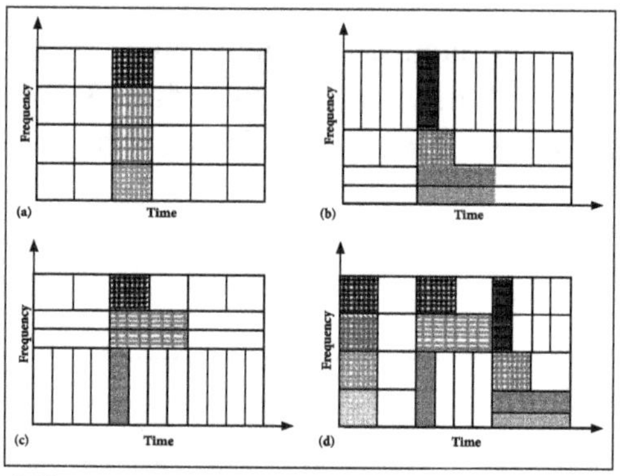

The time frequency tiling for different time frequency transforms: (a) The STFT, (b) The wavelet transform, (c) An example of general tiling and (d) An example of the time varying tiling.

In conventional time frequency transforms, the underlying basis functions are fixed in time and define a specific tiling of the time frequency plane. The term time frequency tile of a particular basis function is meant to designate the region in the plane that contains most of that function's energy. The STFT and the wavelet transform are just two of many possible tilings of the time frequency plane. These two are illustrated in below figure (a) and (b), respectively.

In these figures, the rectangular representation for a tile is purely symbolic, since no function can have compact support in both time and frequency. Other arbitrary filings of the time frequency plane are possible. In discrete domain, linear time frequency transforms can be implemented in the form of filter bank structures.

It is known that the time frequency energy distribution of signals often changes with time. Thus, in this sense, the conventional linear time frequency transform paradigm is fundamentally mismatched to many signals of interest. A more flexible and accurate approach is obtained if the basic functions of the transform are allowed to adapt to the signal properties. An example of such a time varying tiling is shown in figure (d). In this, the time frequency tiling of the transform can be changed from good frequency localization to good time localization and vice versa.

Time varying filter banks provide such flexible and adaptive time frequency tilings. The ideas underlying this method were later developed and extended to a more general case in which it was also shown that the number of frequency bands could also be made adaptive.

4.6.1 Short Time Fourier Transform

The Short Time Fourier Transform (STFT) or Short Term Fourier Transform is a powerful general purpose tool for audio signal processing. It defines a particularly useful class of time frequency distributions, which specify complex amplitude versus time and frequency for any signal.

We are primarily concerned here with tuning the STFT parameters for the following applications:

- Approximating the time frequency analysis performed by the ear for the purpose of spectral display.
- Measuring the model parameters in a short time spectrum.

In the first case, applications of audio spectral display go beyond merely looking at the spectrum. They also provide a basis for audio signal processing tasks intended to imitate human perception, such as auditory scene recognition or automatic transcription of music.

Examples of the second case include estimating the decay time versus frequency for vibrating strings and body resonances or measuring as precisely as possible the fundamental frequency of a periodic signal based on tracking its many harmonics in the STFT.

An interesting example for which cases 1 and 2 normally coincide is pitch detection and fundamental frequency estimation. Here, fundamental frequency is defined as the lowest frequency present in a series of harmonic overtones, whereas pitch is defined as the perceived fundamental frequency. For example, perceived pitch can be measured by comparing it with a harmonic reference tone such as a saw tooth waveform.

Thus, the pitch of a saw tooth waveform is its fundamental frequency. When harmonics are stretched, they become slightly inharmonic; pitch perception corresponds to a compromise fundamental frequency, the harmonics of which best fit the most audible overtones in some sense.

Mathematical Definition of STFT

The usual mathematical definition of the STFT is,

$$X_m(\omega) = \sum_{n=-\infty}^{\infty} x(n)w(n-mR)e^{-j\omega n}$$

$$= \mathrm{DTFT}_{\omega}\left(x \cdot \mathrm{SHIFT}_{mR}(w)\right), \qquad \ldots(1)$$

Where,

$x(n) =$ Input signal at time n.

$w(n) =$ Length M window function, example, Hamming.

$X_m(\omega) =$ DTFT of windowed data centered about time mR.

R = Hop size, in samples, between successive DTFTs.

If the window w(n) has the Constant OverLap Add (COLA) property at hop-size R, (i.e.) if,

$$\sum_{m=-\infty}^{\infty} w(n-mR) = 1,\ \forall n \in Z\ \left(w \in \text{COLA}\ (R)\right) \qquad ...(2)$$

Then the sum of the successive DTFTs over time equals the DTFT of the whole signal $X(\omega)$:

$$\sum_{m=-\infty}^{\infty} X\ (\omega)\ \sum_{m-\infty}^{\infty} \sum_{n=-\infty}^{\infty} x(n)w(n-mR)e^{-j\omega n}$$

$$= \sum_{n=-\infty}^{\infty} x(n)e^{-j\omega n} \underbrace{\sum_{m=-\infty}^{\infty} w(n-mR)}_{1 \text{ if } w \in \text{COLA } (R)}$$

$$= \sum_{n=-\infty}^{\infty} x(n)e^{-j\omega n}$$

$$\triangleq \text{DTFT}_\omega(x) = X(\omega)$$

We will say that windows satisfying $\sum_m \omega(n-mR) = 1$ for all n ϵ Z are said to be COLA(R). For example, the length M rectangular window is clearly COLA(M). The Bartlett window and all windows in the generalized Hamming family are COLA(M/2), (i.e.) 50% overlap, when the endpoints are handled correctly. A COLA(M/2) example is depicted in the below figure. Any window that is COLA(R) is also COLA(R/k), for k = 2, 3, 4, ..., R, provided R/k is an integer.

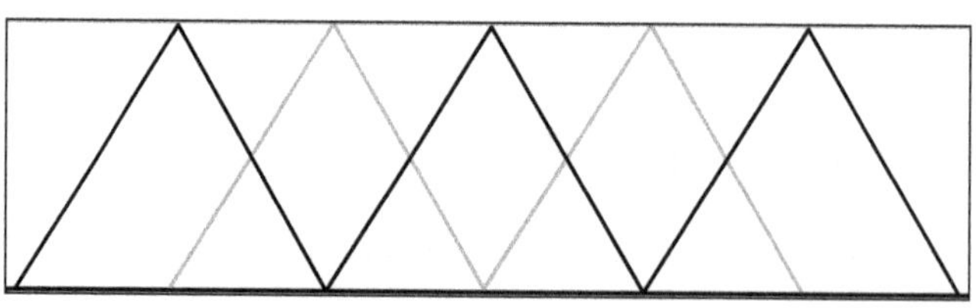

50% overlap-add for the Bartlett window.

When using the Short Time Fourier Transform for signal processing, the COLA requirement is important for avoiding artifacts. For usage as a spectrum analyzer for measurement and display, the COLA requirement can be relaxed, as doing only means that we are not weighting all information equally in our analysis.

For example, if we use 50% overlap with the Blackman window in a short time spectrum analysis over time, the result looks fine, however, in such a case, data falling near the edges of the window will have a slightly muted impact on the results, relative to the data falling near the window center, because the Blackman window is not COLA at 50% overlap.

4.6.2 Spectrogram

The concept of spectrogram also known as the sonogram is very close to that of a spectrum. A spectrum describes the sound signal in terms of energy spread over its frequency components. A spectrogram does exactly the same, but also takes another factor into account, (i.e.) time. While spectra provide stationary information about frequency and energy, spectrograms also display the changes that occur over time.

Speech is a continuous flow, phones are not discrete or distinct from each other, but they merge into one another and spectrograms assess and visualize this continuity.

In spectrograms, time is displayed on the horizontal axis and frequency is displayed on the vertical axis. The amplitude of the frequency components is expressed by means of the degree of blackness. As formants are frequency regions with a high energy, in the spectrogram they are displayed as dark bands.

The instrument that performs the analog spectrogram analysis is a spectrograph or sonograph. The core of the spectrogram is a tunable band pass filter, which performs its analyses on succeeding time portions of the signal, at regular time intervals, until the whole frequency range has been analyzed.

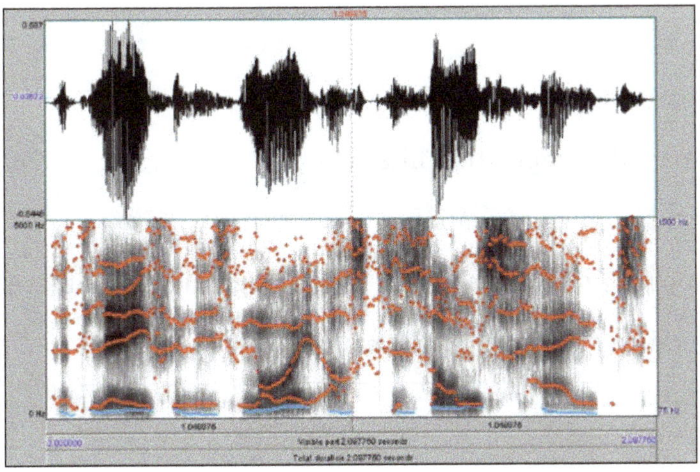

The pitch contour (blue line) represents octave errors, which can be avoided by adjusting the setting parameters.

The degree of blackness is proportional to the amount of energy in that frequency regions. Formants are displayed as black bands. The concentration of energy at the lower end of the spectrogram represents F0 and it may be difficult to distinguish it from the first formant, especially in closed vowels. Formants are more evident in vowels, as they are voiced and there is no anti-resonance effect. The vertical bands at regular interval distance in the spectrum are the glottal pulses.

Spectrograms are obtained by analyzing the spectra of the waveform at regular time intervals. They are estimated by means of Fourier transform.

4.6.3 Wavelet Signal Decomposition

The wavelet decomposition results in levels of approximated and detailed co-efficients. The algorithm of wavelet signal decomposition is illustrated in figure (1). Reconstruction of the signal from the wavelet transform and post processing, the algorithm is shown in figure (2). This multi-resolution analysis enables us to analyze the signal in different frequency bands, therefore, we could observe any transient in time domain as well as in the frequency domain.

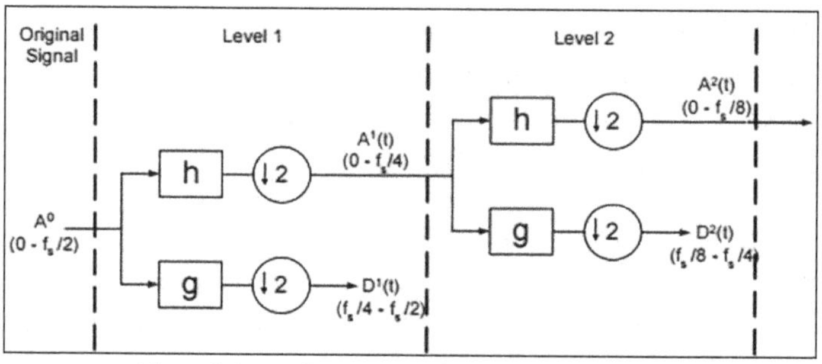

(1) Multi-resolution wavelet decomposition.

Here,

h = Low-pass decomposition filter.

g = High-pass decomposition filter.

⑫ = Down sampling operation.

$A^1(t), A^2(t) =$ Approximated co-efficient of the original signal at levels 1, 2, etc.

$D^1(t), D^2(t) =$ Detailed co-efficient at levels 1, 2.

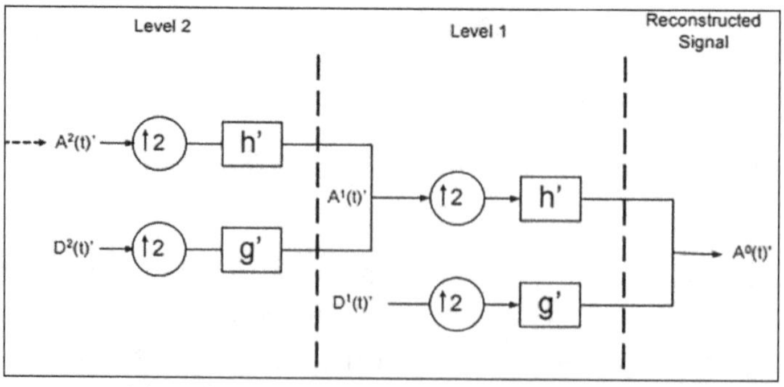

(2) Multi-resolution wavelet reconstruction.

Here,

h' = Low pass reconstruction filter.

g' = High pass reconstruction filter.

⑫ = Up sampling operation.

$A^1(t)', A^2(t)'$ = Processed or non-processed approximated co-efficient of the original signal at levels 1, 2, etc.

$D^1(t)', D^2(t)'$ = Processed or non-processed detailed co-efficient at levels 1, 2.

Level of Decomposition

Wavelet transform constitute of different levels. The maximum level to apply the wavelet transform depends on how many data points are contained in a data set, since there is a down sampling by two operations from one level to the next one. One factor that affects the number of level that can be reached to achieve satisfactory noise removal results is the signal-to-noise ratio in the original signal.

Generally, the measured signals from the Piezoelectric (PZT) sensor have higher SNR than that of the measured signal from fiber optic sensors. So to process the PZT data, we need more level of wavelet transform to remove most of its noise. For the fiber optic sensor data, we could only go up to four or five levels, otherwise we would remove much of the PD signal, therefore the PD spikes would not be captured.

4.7 Fourier, Laplace and Z Transformations

A function $F(\omega)$ is called the Fourier transform of f(x).

If, $F(\omega): \mathcal{F}\{f(t)\} := \int_{-\infty}^{\infty} e^{-i\omega t} f(t)\,dt$ exist. $\mathcal{F}^{-1}\{F(\omega)\} := \frac{1}{2\pi}\int_{-\infty}^{\infty} e^{-i\omega t} F(\omega)\,d\omega$ is called the inverse Fourier transform of $F(\omega)$.

The Fourier transform of f is therefore a function F{f(t)} of the new variable ω. This function, evaluated at ω, is $F(\omega)$. The constants 1 and $1/2\pi$ preceding the integral signs could be replaced by any two constants whose product is $1/2\pi$.

The Fourier transform of f given by,

$$f(t) := \begin{cases} 1 & |t| < 1 \\ 0 & |t| > 1 \end{cases}$$

$$\mathcal{F}\{f(t)\} = \int_{-1}^{1} e^{-i\omega t} dt = \begin{Bmatrix} 2\frac{\sin\omega}{\omega} & \omega \neq 0 \\ 2 & \omega = 0 \end{Bmatrix}$$

The inverse Fourier transform computes to,

$$\frac{1}{\pi}\int_{-\infty}^{\infty} e^{i\omega x}\frac{\sin\omega}{\omega}d\omega = \frac{2}{\pi}\int_{0}^{\infty}\cos\omega x\frac{\sin\omega}{\omega}d\omega$$

$$= \frac{1}{\pi}\int_{0}^{\infty}\frac{\sin\omega(x+1)}{\omega}d\omega - \frac{1}{\pi}\int_{0}^{\infty}\frac{\sin\omega(x-1)}{\omega}d\omega$$

$$= \begin{Bmatrix} 1 & |x|<1 \\ 1/2 & |x|=1 \\ 0 & |x|>1 \end{Bmatrix}$$

Properties

1. Linearity property:

If $F_1(s)$ and $F_2(s)$ be the Fourier transform of $f_1(x)$ and $f_2(x)$ respectively, then

$$F[af_1(x) + bf_2(x)] = aF_1(s) + bF_2(s)$$

Where a and b are constants.

Proof:

$$\text{L.H.S} = \int_{-\infty}^{\infty}[af_1(t) + bf_2(t)] \,.\, e^{ist}\,dt$$

$$= a\int_{-\infty}^{\infty} f_1(t)\,.\,e^{ist}\,dt + b\int_{-\infty}^{\infty} f_2(t)\,.e^{ist}\,dt$$

$$= a\,F_1(s) + b\,F_2(s) = \text{R.H.S.}$$

2. Change of scale property:

If $F[(t)] = F(s)$, then $F[f(at)] = (1/a)\ F(s/a)$, $a > 0$.

Proof:

$$F[f(at)] = \int_{-\infty}^{\infty} f(at).e^{ist}\,dt$$

$$= \frac{1}{a}\int_{-\infty}^{\infty} f(p).e^{i\frac{s}{a}p}dp. \qquad \text{(Taking } at = p \text{ and } adt = dp\text{)}$$

$$= \frac{1}{a}.F(s/a)$$

From this change of scale property, it is evident that if the width of a function is decreased while its height is kept constant, then its Fourier transform wider and shorter. If its width is increased, its transform becomes narrower and taller.

A similar frequency scaling property is given by,

$$F\left[\frac{1}{a}.f(t/a)\right]=F(as).$$

3. Shifting property:

If $F\{f(t)\}=F(s)$ and a is real constant, then

$$F\{f(t-a)\} = e^{isa}.F(s)$$

Proof:

$$F\{f(t-a)\}=\int_{-\infty}^{\infty}f(1-a)e^{ist}\,dt$$

$$=e^{isa}\int_{-\infty}^{\infty}f(t-a)e^{ist(t-a)}.dt$$

$$=e^{isa}.\int_{-\infty}^{\infty}f(p)e^{isp}\,dp \qquad (\text{taking } t-a=p \text{ and } dt=dp)$$

$$=e^{isa}.F(s).$$

This (time) shifting property states that the Fourier transform of a shifted function is just the transform of the unshifted function multiplied by an exponential factor having a linear phase. Likewise, the frequency shifting property states that if F(s) is shifted by a constant 'a', its inverse transform is multiplied by e^{-ita}.

$$F\{f(t).e^{-ita}\}=F\{s-a\}$$

4. Modulation property:

If $F\{f(t)\}=F(s)$, then $F\{f(t)\ \cos wt\}=(1/2)[F(s-w)+F(s+w)]$.

Proof:

$$F[f(t).\cos wt]=\int_{-\infty}^{\infty}f(t).\cos wt.e^{ist}.dt$$

$$=\int_{-\infty}^{\infty}f(t).\left(\frac{e^{iwt}+e^{iwt}}{2}\right).e^{ist}\,dt$$

$$=\frac{1}{2}.\left[\int_{-\infty}^{\infty}f(t).e^{i(x+w)t}\,dt+\int_{-\infty}^{\infty}f(t).e^{i(x-w)t}\,dt\right]$$

$$=\frac{1}{2}[F(s+w)+F(s-w)].$$

This result has application in radio and television where the harmonic carrier wave is modulated by an envelope.

5. Convolution property:

The convolution of two functions f(t) and g(t) in (-∞, ∞) is defined as

$$f(t)*g(t)=\int_{-\infty}^{\infty} f(t)\,.\,g(x-t)dt].$$

Let, $F\{f(t)\}=F_1(s)$ and $F\{g(t)\}=F_2(s)$, then,

$$F[f(t)*g(t)]=F_1(s).\,F_2(s)$$

Proof:

$$F[f(t)*g(t)]=\int_{-\infty}^{\infty}\{f(t)*g(t)\}.e^{ist}\,dt$$
$$=\int_{-\infty}^{\infty}\left(\int_{-\infty}^{\infty} f(u)g(t-u)du\right)e^{ist}\,dt.$$
$$=\int_{-\infty}^{\infty} f(u)e^{isu}\left(\int_{-\infty}^{\infty} g(t-u).e^{is(t-u)}d(t-u)\right)du$$

On changing the order of integration,

$$=\int_{-\infty}^{\infty} f(u).e^{isu}.F_2(s)du=\left(.\int_{-\infty}^{\infty} f(u).e^{isu}\,du\right).F_2(s)$$
$$=F_1(s).F_2(s)$$

6. Correlation property:

The correlation integral, like the convolution integral is important in theoretical and practical applications.

The correlation integral is defined as,

$h(t)=\int_{-\infty}^{\infty} f(u)g(t+u)du$ and like the convolution property, it forms a transform pair given by,

$$F\{h(t)\}=F(s)\,.\,G^*(s)$$

If f(t) and g(t) are the same function, the integral above is normally called the autocorrelation function and the cross correlation if they differ. The Fourier transform pair for the auto correlation is simply,

$$F\left\{\int_{-\infty}^{\infty} f(u).f(t-u)du\right\}=|F(s)|^2$$

4.7.1 Laplace

The Laplace transform is particularly useful in solving linear ordinary differential equations such as those arising in the analysis of electronic circuits.

The (unilateral) Laplace transform L is defined by,

$$\mathcal{L}_t[f(t)](s) \equiv \int_0^\infty f(t)e^{-st}\,dt,$$

Where $f(t)$ is defined for $t \geq 0$

S. No.	Input	Laplace Transform
1.	1	$1/S$
2.	e^{at}	$1/(s-a)$
3.	$t^n, n = 1, 2, 3, \ldots$	$\frac{n!}{S^{n+1}}$
4.	$t^p, p > -1$	$\frac{\Gamma(p+1)}{s^{p+1}}$
5.	$\sin(at)$	$\frac{a}{s^2 + a^2}$
6.	$\cos(at)$	$\frac{s}{s^2 + a^2}$
7.	$t\sin(at)$	$\frac{2as}{(s^2 + a^2)^2}$
8.	$t\cos(at)$	$\frac{s^2 - a^2}{(s^2 + a^2)^2}$
9.	$\sin(at) - at\cos(at)$	$\frac{2a^3}{(s^2 + a^2)^2}$
10.	$\sin(at) + at\cos(at)$	$\frac{2as^2}{(s^2 + a^2)^2}$
11.	$\cos(at) - at\sin(at)$	$\frac{s(s^2 - a^2)}{(s^2 + a^2)^2}$
12.	$\cos(at) + at\sin(at)$	$\frac{s(s^2 + 3a^2)}{(s^2 + a^2)}$
13.	$\sin(at + b)$	$\frac{s\sin(b) + a\cos(b)}{s^2 + a^2}$

14.	$\cos(at+b)$	$\frac{s\cos(b)-a\sin(b)}{s^2+a^2}$
15.	$\sinh(at)$	$a/(s^2-a^2)$
16.	$\cosh(at)$	$\frac{s}{s^2-a^2}$
16.	$e^{at}\sin(bt)$	$\frac{b}{(s-a)^2+b^2}$
17.	$e^{at}\cos(bt)$	$\frac{s-a}{(s-a)^2+b^2}$
18.	$e^{at}\sinh(bt)$	$\frac{b}{(s-a)^2-b^2}$
19.	$e^{at}\cosh(bt)$	$\frac{s-a}{(s-a)^2-b^2}$
20.	$t^n e^{at}$, $n=1, 2, 3, \ldots.$	$\frac{n!}{(s-a)^{n+1}}$
21.	$f(ct)$	$\frac{1}{c}F\left(\frac{s}{c}\right)$
22.	$u_c(t)=u(t-c)$	e^{-cs}/s
23.	$\delta(t-c)$	e^{-cs}
24.	$u_c(t)f(t-c)$	$e^{-cs}F(s)$
25.	$u_c(t)g(t)$	$e^{-cs}\,\varepsilon\{g(t+c)\}$
26.	$e^{ct}f(t)$	$F(s-c)$
27.	$t^n f(t)$, $n=1, 2, 3, \ldots.$	$(-1)^n F^{(n)}(s)$
28.	$\frac{1}{t}f(t)$	$\int_s^\infty F(u)du$
29.	$\int_o^t f(v)dv$	$F(s)/s$
30.	$\int_o^t f(t-\tau)g(\tau)d\tau$	$F(s)G(s)$

31.	$f(t+T)=f(t)$	$\frac{\int_0^T e^{-st}f(t)dt}{1-e^{-st}}$
32.	$f'(t)$	$sF(s) - f(0)$
33.	$f^n(t)$	$s^2F(s) - sf(0) - f'(0)$
34.	$f^{(n)}(t)$	$s^nF(s) - s^{n-1}f(0) - s^{n-2}f'(0).....-sf^{(n-2)}(0) - f^{(n-1)}(0)$

The Laplace transform is particularly useful in solving linear ordinary differential equations such as those arising in the analysis of electronic circuits.

The (unilateral) Laplace transform L is defined by,

$$\mathcal{L}_t[f(t)](s) \equiv \int_0^\infty f(t)e^{-st}\, d\, t,$$

Where f(t) is defined for $t \geq 0$.

f(t)	F(s) = L(f(t))	
1	1/s	$s > 0$
tn	$\frac{n!}{s^{n+1}}$	$s > 0, n = 0, 1, 2,...$
ta	$\frac{\Gamma(a+1)}{s^{a+1}}$	$s > 0$
eat	$\frac{1}{s-a}$	$s > a$
cos h(kt)	$\frac{s}{s^2-k^2}$	$s > \|k\|$
sin h(kt)	$\frac{k}{s^2-k^2}$	$s > \|k\|$
Cos(kt)	$\frac{s}{s^2+k^2}$	$s > 0$
sin(kt)	$\frac{k}{s^2+k^2}$	$s > 0$
H(t -a)	$\frac{e^{-as}}{s}$	$s > 0, a > 0$

First Shifting Theorem

If f(t) is a causal function with Laplace transform F(s), (i.e.) L{f(t)} = F(s), then we shall see, the Laplace transform of $e^{-at}f(t)$.

Where, a is a given constant, can easily be found in terms of F(s). Using the definition of the Laplace transform,

$$n - m = P$$
$$X = X(t)$$
$$Y = X(t - \tau)$$

But if,

$$F(s) = \mathcal{L}\{f(t)\} = \int_0^\infty e^{-st} f(t)dt$$

Then by simply replacing s by s + a on both the sides.

$$F(s+a) = \int_0^\infty e^{-(s+a)t} f(t)dt$$

That is, the parameter s is shifted to the value s + a. We have then the statement of the first shift theorem.

For example, we already know that,

$$\mathcal{L}\{t^3 u(t)\} = \frac{6}{s^4}$$

And so by the first shift theorem,

$$\mathcal{L}\{e^{-2t} t^3 u(t)\} = \frac{6}{(s+2)^4}$$

Transform of an Integral: Theorem

Let F = L[f] where f is any piecewise continuous function on (0, ∞) of exponential order s_0 and let s_1 be any positive value greater than or equal to s_0. Then,

$$\int_0^t f(\tau)d\tau$$

Is a continuous function of ton (0,∞) of exponential order s_1 and, for each positive s greater than s_0,

$$\mathcal{L}\left[\int_0^I f(\tau)d\tau\right]\Big|_s = \frac{F(s)}{s}$$

Problems

1. Let us determine the Laplace transform of,

$$f(t)=\begin{cases} 1, & 0\le t<2, \\ t-2, & 2\le t. \end{cases}$$

Solution:

Given:

We do this by definition.

$$F(s)=\int_0^\infty e^{-st}f(t)dt=\int_0^2 e^{-st}dt+\int_2^\infty (t-2)e^{-st}dt$$

$$=\frac{1}{-s}e^{-st}\Big|_{t=0}^{1}+(t-2)\frac{1}{-s}e^{-st}\Big|_{t=2}^{\infty}-\int_2^A\frac{1}{-s}e^{-st}dt$$

$$=\frac{1}{-s}\left(e^{-2s}-1\right)+(0-0)+\frac{1}{s}\frac{1}{-s}e^{-st}\Big|_{t=2}^{\infty}=\frac{1}{-s}\left(e^{-2s}-1\right)+\frac{1}{s^2}e^{-2s}$$

2. Let us determine $L[t\sin at]$.

Solution:

Given:

If, $L[f(t)]=F(s)$ then $L[t\,f(t)]=-\frac{d}{ds}F(s)=-F'(s)$

We know that,

$$L[t\sin at]=-\frac{d}{ds}L[\sin at]$$

$$=-\frac{d}{ds}\left[\frac{a}{s^2+a^2}\right]$$

$$=-\left(\frac{(s^2+a^2)(0)-a(2s)}{(s^2+a^2)}\right)$$

$$=-\left(\frac{-a(2s)}{(s^2+a^2)^2}\right)$$

$$\therefore L[t\sin at]=\frac{2as}{(s^2+a^2)^2}$$

3. Let us show that $\int_0^\infty e^{-t} t \cos t dt = 0$.

Solution:

Given:

$$\int_0^\infty e^{-t} t \cos t dt = 0$$

$$\int_0 e^{-t} t \cos t \, dt = \left[L[t \cos t]\right]_{s=1}$$

$$= \left[-\frac{d}{ds} L(\cos t)\right]_{s=1}$$

$$= \left[-\frac{d}{ds}\left(\frac{s}{s^2+1}\right)\right]_{s=1}$$

$$= \left[-\left[\frac{(s^2+1)(1)-s(2s)}{(s^2+1)^2}\right]\right]_{s=1}$$

$$= \left[-\left[\frac{s^2+1-2s^2}{(s^2+1)^2}\right]\right]_{s=1} = \left[-\left[\frac{1-s^2}{(s^2+1)^2}\right]\right]_{s=1} = [-0]$$

$$\therefore \int_0^\infty e^{-t} t \cos t \, dt = 0$$

4. We shall calculate $L\left[\frac{\sin 3t \cos t}{t}\right]$.

Solution:

Given:

$$L\left[\frac{\sin 3t \cos t}{t}\right]$$

$$L\left[\frac{\sin 3t \cos t}{t}\right] = L\left[\frac{\sin 4t + \sin 2t}{2t}\right]$$

$$= \frac{1}{2} L\left[\frac{\sin 4t + \sin 2t}{t}\right]$$

$$=\frac{1}{2}\left[L\left[\frac{\sin 4t}{t}\right]+L\left[\frac{\sin 2t}{t}\right]\right]$$

$$=\frac{1}{2}\left(\cot^{-1}\left(\frac{s}{4}\right)+\cot^{-1}\left(\frac{s}{2}\right)\right) \qquad \left(\because L\left[\frac{\sin at}{t}\right]=\cot^{-1}\left(\frac{s}{a}\right)\right)$$

$$\therefore L\left[\frac{\sin 3t\cos t}{t}\right]=\frac{1}{2}\left(\cot^{-1}\left(\frac{s}{4}\right)+\cot^{-1}\left(\frac{s}{2}\right)\right)$$

5. We shall calculate $L\left[\dfrac{\cos at-\cos bt}{t}\right]$.

Solution:

Given:

$$L\left[\frac{\cos at-\cos bt}{t}\right]$$

$$L\left[\frac{\cos at-\cos bt}{t}\right]=\int_s^\infty L[\cos at-\cos bt]ds$$

$$=\int_s^\infty\left[\frac{s}{s^2+a^2}-\frac{s}{s^2+b^2}\right]ds$$

$$=\frac{1}{2}\left[\log\left(s^2+a^2\right)-\log\left(s^2+b^2\right)\right]_s^\infty$$

$$=\frac{1}{2}\left[\log\frac{s^2+a^2}{s^2+b^2}\right]_s^\infty$$

$$=\frac{1}{2}\left[0-\log\frac{s^2+a^2}{s^2+b^2}\right] \qquad (\because \log 1=0)$$

$$\therefore L\left[\frac{\cos at-\cos bt}{t}\right]=\frac{1}{2}\left[\log\frac{s^2+a^2}{s^2+b^2}\right]$$

6. Let us solve, by Laplace transform method, the equation.

Given:

$$\frac{d^2y}{dt^2}+2\frac{dy}{dt}+5y=e^{+}\sin t,\ y(0)=0,\ y(0)=1.$$

Solution:

$$y''(t)+2y'(t)+5y(t)=e^{-t}\sin t \qquad \dots (1)$$

Taking L.T on both sides,

$$L[y''(t)]+2L[y'(t)]+5L[y(t)]=L[e^{-t}\sin t].$$

$$\Rightarrow[S^2L[y(t)]-sy(\not{0})-y(0)]+2[SL[y(t)]-y(\not{0})]+5L[y(t)]$$

$$=\left[\frac{1}{s^2+1}\right]_{s\to s+1}$$

Given y(0) = 0y' (0) = 1 and let $L[y(t)]=\bar{y}$

$$\Rightarrow s^2\bar{y}-1+2[s\bar{y}]+5\bar{y}=\frac{1}{(s+1)^2+1}$$

$$\bar{y}[s^2+2s+5]=\frac{1}{(s+1)^2+1}+1=\frac{1+(s+1)^2+1}{(s+1)^2+1}$$

$$\bar{y}(s^2+2s+5)=\frac{1+s^2+2s+1+1}{(s+1)^2+1}=\frac{s^2+2s+3}{s^2+2s+2}$$

$$\bar{y}=\frac{s^2+2s+3}{(s^2+2s+2)(s^2+2s+5)}.$$

Let $s^2+2s=x$

$$\therefore\bar{y}=\frac{x+3}{(x+2)(x+5)}=\frac{A}{x+2}+\frac{B}{x+5}$$

$$x+3=A(x+5)+B(x+2)$$

$$x=-2\Rightarrow 1=3A\quad A=\frac{1}{3}$$

$$x=-5\Rightarrow -2=-3B\quad B=\frac{2}{3}.$$

$$\therefore\bar{y}=\frac{1/3}{x+2}+\frac{2/3}{x+5}\quad\text{where } x=s^2+2s$$

$$\therefore y=\frac{1}{3}\frac{1}{s^2+2s+2}+\frac{2}{3}\frac{1}{s^2+2s+5}$$

$$L[y(t)]=\frac{1}{3}\frac{1}{(S+1)^2+1}+\frac{2}{3}\frac{1}{(s+1)^2+2^2}$$

$$y(t)=\frac{1}{3}L^{-1}\left[\frac{1}{(s+1)^2+1}\right]+\frac{2}{3}L^{-1}\left[\frac{1}{(s+1)^2+2^2}\right]$$

$$=\frac{1}{3}e^{-t}L^{-1}\left(\frac{1}{s^2+1}\right)+\frac{2}{3}e^{-t}L^{-1}\left(\frac{1}{s^2+2^2}\right)$$

$$=\frac{1}{3}e^{-t}\sin t+\frac{2}{3}e^{-t}.\frac{\sin 2t}{2}.$$

$$y(t)=\frac{e^{-t}}{3}(\sin t+\sin 2t)$$

7. Using Laplace transform let us solve the differential equation $y"-3y'-4y=2e^{-t}$ with y(0) = 1 = y (0).

Solution:

Given:

$$y"(t)-3y'(t)-4y\ (t)=2e^{-t} \qquad ...(1)$$

Consider,

$$L[y(t)]=F(s)$$

$$L[y'\ (0]=s\,F(s)-y(0)$$

$$L[y"\ (t)]=s^2\ F(s)-sy(0)-y'\ (0)$$

Taking Laplace transforms on both sides of (1),

$$L[y"(t)]-3\,L[y'(t)]-4L[y(t)]=2L[e^{-t}]$$

$$[s^2\,F(s)-sy(0)-y(0)]-3[s\,F(s)-y(0)]-4\,F(s)=\frac{2}{s+1}$$

$$s^2F(s)-s-1-3s\ F(s)+3-4F(s)=\frac{2}{(s+1)}$$

$$(s^2-3s-4)\ F(s)=\frac{2}{(s+1)}+s-2$$

$$=\frac{s^2-s}{(s+1)}$$

$$L\left[y(t)\right]=F(s)=\frac{s^2-s}{(s+1)^2(s-4)}$$

i.e., $y(t)=L^{-1}\left[\frac{s^2-s}{(s+1)^2(s-4)}\right]$ ...(1)

Consider,

$$\frac{s^2-s}{(s+1)^2(s-4)}=\frac{A}{(s+1)}+\frac{B}{(s+1)^2}+\frac{C}{s-4} \quad ...(2)$$

$$s^2-s=A(s+1)(s-4)+B(s-4)+C(s+1)^2 \quad ...(3)$$

Put s = - 1

$$2=-5B \Rightarrow B=-2/5$$

Put s = 4

$$12=25C \Rightarrow C=12/25$$

Put s = 0

$$0=-4A-4B+C$$

$$4A=C-4B=\frac{12}{25}+\frac{8}{5}=\frac{52}{25}$$

$$\Rightarrow A=\frac{13}{25}$$

$$\frac{s^2-s}{(s+1)^2(s-4)}=\frac{13}{25(s+1)}-\frac{2}{5(s+1)^2}+\frac{12}{25(s-4)} \quad ...(4)$$

Taking L^{-1} on the both sides,

$$L^{-1}\left[\frac{s^2-s}{(s+1)^2(s-4)}\right]=\frac{13}{25}L^{-1}\left[\frac{1}{s+1}\right]-\frac{2}{5}L^{-1}\left[\frac{1}{(s+1)^2}\right]+\frac{12}{25}L^{-1}\left[\frac{1}{s-4}\right]$$

i.e., $y(t)=\frac{13}{25}e^{-t}-\frac{2}{5}e^{-t}L^{-1}\left[\frac{1}{s^2}\right]+\frac{12}{25}e^{4t}$

$$y(t)=\frac{13}{25}e^{-t}-\frac{2te^{-t}}{5}+\frac{12}{25}e^{4t}.$$

8. We shall evaluate $\int_0^\infty t\, e^{-2t} \cos t\, dt$ using Laplace Transform.

Solution:

Given:

$$\text{Let } I = \int_0^\infty t\, e^{-2t} \cos t\, dt = \int_0^\infty e^{-2t} (t \cos t) dt$$

$$= \int_0^\infty e^{-st} (t \cos t) dt \quad \text{where } s = 2$$

$$= L[t \cos t];\ s = 2$$

$$= -\frac{d}{ds} L[\cos t];\ s = 2$$

$$= -\frac{d}{ds}\left[\frac{s}{s^2+1}\right];\ s = 2$$

$$= -\left[\frac{(s^2+1)(1) - s(2s)}{(s^2+1)^2}\right];\ s = 2$$

$$= -\left[\frac{s^2+1-2s^2}{(s^2+1)^2}\right];\ s = 2$$

$$= -\left[\frac{-s^2+1}{(s^2+1)^2}\right];\ s = 2$$

$$= \frac{s^2-1}{(s^2+1)^2};\ s = 2$$

$$= \frac{4-1}{(4+1)^2} = \frac{3}{25}$$

9. Let us determine $L\left[t^2 e^{-3t} \sin 2t\right]$.

Solution:

Given:

$$L\left[t^2 e^{-3t} \sin 2t\right] = L\left[t^2 \sin 2t\right]_{s \to S+3} \qquad \text{...(1)}$$

$$L\left[t^2 \sin 2t\right] = \frac{d^2}{ds^2} L[\sin 2t]$$

Now

$$= \frac{d^2}{ds^2} \frac{2}{s^2+4}$$

$$= 2\frac{d}{ds}\left[\frac{d}{ds}\left(s^2+4\right)^{-1}\right]$$

$$= 2\frac{d}{ds}\left[-2s\left(s^2+4\right)^{-2}\right]$$

$$= -4\frac{d}{ds}\frac{s}{\left(s^2+4\right)^2}$$

$$= -4\left[\frac{\left(s^2+4\right)^2(1) - s(2)\left(s^2+4\right)(2s)}{\left(s^2+4\right)^4}\right]$$

$$= -4\left(s^2+4\right)\frac{\left[S^2+4-4s^2\right]}{\left(s^2+4\right)^4}$$

$$= \frac{4\left(3s^2-4\right)}{\left(s^2+4\right)^3} \qquad \text{...(2)}$$

Substitute (2) in (1)

$$L\left[t^2 e^{-3t} \sin 2t\right] = \left\{\frac{12s^2-16}{\left(s^2+4\right)^3}\right\}_{s\to s+3}$$

$$= \frac{12(s+3)^2-16}{\left[(s+3)^2+4\right]^3}.$$

10. Here we shall determine $L\left[\frac{\cos at - \cos bt}{t}\right]$.

Solution:

Given:

$$L\left[\frac{\cos at - \cos bt}{t}\right] = \int_s^\infty L[\cos at - \cos bt]ds$$

$$= \int_{s}^{\infty} \left[L(\cos at) - L(\cos bt) \right] ds$$

$$= \int_{0}^{\infty} \left[\frac{s}{s^2 + a^2} - \frac{s}{s^2 + b^2} \right] ds$$

$$= \left[\frac{1}{2} \log\left(s^2 + a^2\right) - \frac{1}{2} \log\left(s^2 + b^2\right) \right]_s^{\infty}$$

$$= \frac{1}{2} \left[\log \frac{\left(s^2 + a^2\right)}{\left(s^2 + b^2\right)} \right]_s^{\infty}$$

$$S \to \infty \Rightarrow \frac{1}{s} \to 0$$

$$= \frac{1}{2} \left[\log \frac{\left(1 + \frac{a_2}{b^2}\right)}{\left(1 + \frac{b^2}{s^2}\right)} \right]_s^{\infty}$$

$$= \frac{1}{2} \left[\log 1 - \log \frac{\left(s^2 + a^2\right)}{\left(s^2 + b^2\right)} \right]$$

$$= -\frac{1}{2} \log \frac{\left(s^2 + a^2\right)}{\left(s^2 + b^2\right)}$$

$$L\left[\frac{\cos at - \cos bt}{t} \right] = \frac{1}{2} \log \frac{\left(s^2 + b^2\right)}{\left(s^2 + a^2\right)}$$

4.7.2 Z-transforms

The z-transform is named as a letter of the alphabet rather than a famous mathematician. A method for solving linear constant co-efficient difference equations by Laplace transforms was introduced to by Gardner and Barnes in the early 1940s. They applied their procedure which was based on the jump functions to transmission lines and applications involving Bessel functions. This approach is quite complicated and in a separate attempt to simplify matters, a transform of the sampled signal or sequence was defined in 1947 by W. Hurewicz as which was later denoted in 1952 as a Z-transform by a sampled-data control group at Colombia University.

In any case, z-transform was invented at about the same time as digital computers. However z-transform can be viewed as a mathematical operation which takes a set of points and transforms them into a set of complex numbers. Z-transform is mathematical tool used for the conversion of time domain into frequency domain and is a function of complex valued variable Z. The z-transform of a discrete time signal x(n) is denoted by X(z) and is given as,

$$X(z)=\sum_{n--\infty}^{\infty} x[n]z^{-n}$$

Z-transform is an infinite power series because summation index varies from -∞ to ∞. But it is useful for values of z for which sum is finite.

Advantages of Z-Transform

- The DFT can be determined by evaluating z-transform.
- Z-transform is widely used for analysis and synthesis of digital filter.
- Z-transform is used for linear filtering.
- Z-transform is also used for finding Linear convolution, cross-correlation and auto-correlations of sequences.
- In Z-transform, user can characterize LTI system (stable/unstable, causal/anti-causal) and its response to various signals by placements of pole and zero plot.

Some useful Z transforms are:

$X(z)$	$x(n)=Z^{-1}.\ (X(z))$
$\frac{1}{z-a}$	$a^{n-1}.u(n-1)$
$\frac{1}{(z-a)^2}$	$(n-1)a^{n-2}.u(n-2)$
$\frac{1}{(z-a)^3}$	$\frac{1}{2}(n-1)(n-2)a^{n-3}.u(n-3)$
$\frac{z}{z-a}$	$a^n.u(n)$
$\frac{z^2}{(z-a)^2}$	$(n+1)a^n u(n)$
$\frac{z^3}{(z-a)^3}$	$\frac{1}{2}(n+1)(n+2)a^n.u(n)$
$\frac{z}{(z-a)^2}$	na^{n-1}

Problems

1. Let us find the z-transform for $u(n-1)$.

Solution:

Given:

$$u(n-1)$$

By using right shift property,

$$Z\{u(n-1)\} = z^{-1}.Z\{u(n)\}$$

$$= z^{-1}.\frac{z}{z-1}$$

$$= \frac{1}{z-1}$$

2. Let us find the z-transform for $na^n.u(n)$.

Solution:

Given:

$$na^n\ .\ u(n)$$

We have,

(using damping rule)

$$Z\{u(n)\} = \frac{z}{z-1}$$

$$\therefore\ z\{a^n u(n)\} = \frac{z/a}{(z/a)-1}$$

$$= \frac{z}{z-a} = \frac{1}{1-az^{-1}} = \left(1-az^{-1}\right)^{-1}$$

(Using differentiating property)

$$Z\{na^n.u(n)\} = z^{-1}\frac{d}{dz^{-1}}\left(1-az^{-1}\right)$$

$$= z^{-1}.(-1)(-a)\left(1-az^{-1}\right)^{-2}$$

$$= \frac{az^{-1}}{\left(1-az^{-1}\right)^2}$$

Properties

1. Linearity:

If $Z\{x_1(n)\} = X_1(z),\ Z\{x_2(n)\} = X_2(z)$, then

$$Z\{ax_1(n) + bx_2(n)\} = aX_1(z) + bX_2(z).$$

Proof:

$$Z\{ax_1(n)+bx_2(n)\} = \sum_{n=-\infty}^{\infty}\left[ax_1(n)+b_1x_2(n)\right]Z^{-n}$$

$$= a\sum_{n=-\infty}^{\infty}x_1(n)Z^{-n} + b\sum_{n=-\infty}^{\infty}x_2(n)Z^{-n}$$

$$= aX_1(z)+bX_2(z).$$

2. Time Shifting:

If $Z\{x(n)\} = X(z)$ and the initial conditions for x(n) are zeros, then $Z\{x(n-m)\} = Z^{-m}X(z)$.

Proof:

$$Z\{x(n-m)\} = \sum_{n=-\infty}^{\infty}x(n-m).Z^{-n}$$

Let $n-m=P$

$$n = p+m$$

$$Z\{x(n-m)\} = \sum_{n=-\infty}^{\infty}x(P).Z^{-(P+m)}$$

$$= Z^{-m}.\sum_{n=-\infty}^{\infty}x(P).Z^{-p}$$

$$= Z^{-m}X(z).$$

3. Multiplication of an Exponential Sequence:

If $Z\{x(n)\} = X(z)$, then

$$Z[a^n x(n)\} = X\left(\alpha^{-1}Z\right)$$

Proof:

$$Z\{a^n x(n)\} = \sum_{n=-\infty}^{\infty} a^n x(n) Z^{-n}$$

$$= \sum_{n=-\infty}^{\infty} x(n) \cdot (a^{-1} Z)^{-n}$$

$$= X(a^{-1} Z)$$

4. Time Reversal:

If Z $\{x(n)\} = X(z)$, then Z $\{x(-n)\} = X(z^{-1})$.

Proof:

$$Z\{x-(-n)\} = \sum_{n=-\infty}^{\infty} x(-n) Z^{-n}$$

$$= \sum_{n=-\infty}^{\infty} x(P) \,.\, Z^{P} \quad P = -n$$

$$= \sum_{n=-\infty}^{\infty} x(P) \,.\, (z^{-1})^{-P}$$

$$= X(z^{-1})$$

5. Differentiation of X(z):

If Z $\{x(n)\} = X(z)$, then

$$Z\{nx(n)\} = -Z \frac{d}{dz} X(z)$$

Proof:

$$X(z) = \sum_{n=-\infty}^{\infty} x(n) Z^{-n}$$

Differentiation on both sizes,

$$\frac{d}{dz} X(z) = \sum_{n=-\infty}^{\infty} x(n) \,.\, \frac{d}{dz}(Z^{-n})$$

$$= \sum_{n=-\infty}^{\infty} x(n).(-n).Z^{-n-1}$$

$$= \sum_{n=-\infty}^{\infty} x(n).(-n).Z^{-n}\ Z^{-1}$$

$$= -\frac{1}{Z}.\sum_{n=-\infty}^{\infty} nx(n)Z^{-n}$$

$$\frac{d}{dz}X(z) = -\frac{1}{z}.Z\{n\ x(n)\}$$

$$-Z\frac{d}{dz}X(z) = Z\{nx(n)\}$$

6. Convolution property of Z-transform:

If, $x(n) \Leftrightarrow X(z);$

$y(n) \Leftrightarrow Y(z)$

Then,

$$x(n) * y(n) \Leftrightarrow X(z)Y(z)$$

7. Initial and final value theorem of Z-transform:

- Initial value theorem $x(0) = \lim_{z \to \infty} X(z).$
- Final value theorem $x(\infty) = \lim_{z \to 1} (1 - z^{-1})X(z).$

4.8 Autocorrelation

The power spectrum of a time series x(t) describes how the variance of the data x(t) is distributed over the frequency components into which x(t) may be decomposed. This distribution of the variance may be described either by a measure μ or by a statistical cumulative distribution function S(f) = Power contributed by frequencies from 0 up to f.

Given a band of frequencies [a, b], the amount of variance contributed to x(t) by frequencies lying within the interval [a, b] is given by S(b) - S(a). Then S is called as the spectral distribution function of x.

The spectral density at a frequency f gives the rate of variance contributed by frequencies in the immediate neighbourhood of f to the variance of x per unit frequency.

Auto Correlation Functions

A stationary random process X(t) is said to be Autocorrelation Ergodic if and only if the time autocorrelation function of any sample function x(t) is equal to the statistical autocorrelation function of X(t). (i.e.) $A\left[x(t)x(t+\tau)\right]=E\left[X(t)\,X(t+\tau)\right]$ or $R_{xx}(\tau)=R_{XX}(\tau)$.

Properties of Auto Correlation

1. R(τ) is an even function of τ.

i.e. $R(\tau)=R(-\tau)$

Proof:

$$R(\tau)=E\big(X(t)\times X(t-\tau)\big)$$

$$R(-\tau)=E\big(X(t)\times X(t+\tau)\big)$$

Put $t+\tau=t'$

$$\therefore t=t'-\tau$$

$$\therefore\ R(-\tau)=E(X(t'-\tau)\times X(t'))=E\big(X(t')\times X(t'-\tau)\big)$$

$$\therefore\ R(-\tau)=R(\tau)$$

2. R(τ) is maximum at τ = 0.

i.e. $|R(\tau)|\le R(0)$

Proof:

Cauchy Schwartz inequality is,

$$E(XY)^2\le E(X^2)\times E(Y^2)$$

Put, $X=X(t)$ and $Y=X(t-\tau)$

Then,

$$\left\{\left\{E\big(X(t)X(t-\tau)\big)\right\}^2\right\}\le E\big(X^2(t)\big)\times E\big(X^2(t-\tau)\big)$$

(i.e.), $\{R(\tau)\}^2\le\left[E\{X^2(t)\}\right]^2$

[Since E{X(t)} and Var {X(t)} are constant for a stationary process]

i.e. $\{R(\tau)\}^2 \le \{R(o)\}^2$

Taking square roots on both sides,

$$|R(\tau)| \le R(o)\left[\text{since } R(o) = E\{X^2\ (t)\} \text{ is positive}\right]$$

3. If the autocorrelation function R(τ) of a real stationary process {X(t)} is continuous at τ = 0, it is continuous at every other point.

If R(τ) is the autocorrelation function of a stationary process {X(t)} with no periodic component, then

$\lim_{\tau\to\infty} R(\tau) = \mu$, provided the limit exists.

Proof:

$$R(\tau) = E(X(t) \times X(t-\tau))$$

When τ is very large, X(t) and X(t - τ) are two sample functions (members) of the process {X(t)} observed at a very long interval of time.

Therefore, X(t) and X(t - τ) tend to become independent [X(t) and X(t - τ) may be dependent, when X(t) contains a periodic component, which is not true]

$$\lim_{\tau\to\infty} R(\tau) = E(X(t) \times X(t-\tau)) = \mu^2_x \text{ (since E(X(t)) is constant)}$$

$$\therefore \mu_x = \sqrt{\lim_{\tau\to\infty} R(\tau)}$$

Practical Applications

- Use of auto-correlation function to quantify the effect of noise on a periodic signal. In absence of noise, the auto-correlation function oscillates with a constant amplitude and a maximum of 1. The period of auto-correlation correspond to the period of the signal. In presence of noise, the envelope of the auto-correlation function decreases exponentially. More the noise, faster it decreases. This phenomenon is also called "phase diffusion".
- Autocorrelation analysis is used heavily in fluorescence correlation spectroscopy.
- It is also used in the measurement of optical spectra and in measurement of very-short-duration light pulses produced by lasers, both using optical auto co-relators.

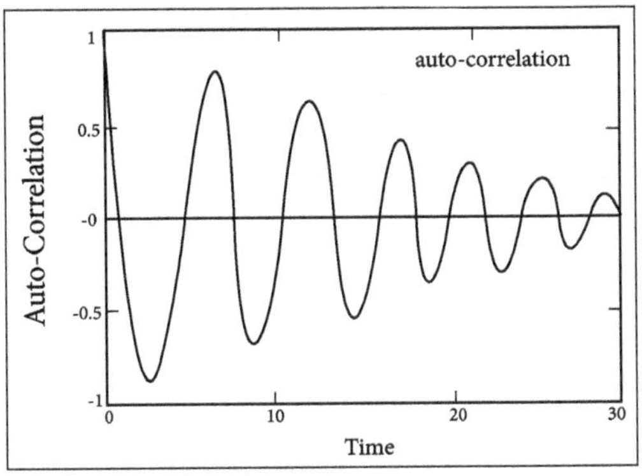

- In optics, normalized autocorrelations and cross-correlations give the degree of coherence of an electromagnetic field.
- In signal processing, autocorrelation can give information about repeating events like musical beats or pulsar frequencies, though it cannot tell the position in time of the beat. It can also be used to estimate the pitch of a musical tone.
- In astrophysics, auto-correlation is used to study and characterize the spatial distribution of galaxies in the Universe and in multi-wavelength observations of Low Mass X-ray Binaries.

Problems

1. Let us check whether the following function are valid auto correlation function:

- $5 \sin n\pi$
- $1/1+9\tau^2$

Solution:

Given:

(i) $R_{xx}(\tau) = 5 \text{ Sin } n\pi$

$$R_{xx}(\tau) = 5 \text{ Sin } n(-\pi) = -5 \text{ Sin } n\pi$$

$R_{xx}(\tau) \neq R_{xx}(-\tau)$, the given function is not an auto correlation function.

(ii) Given $R_{xx}(\tau) = \dfrac{1}{1+9\tau^2}$

$$R_{xx}(-\tau) = \frac{1}{1+9(-\tau)^2} = R_{xx}(\tau)$$

∴ The given function is an auto correlation function.

2. The autocorrelation function of a stationary random process is $R(\tau)=16+\frac{9}{1+6\tau^2}$. Let us calculate the mean and variance of the process.

Solution:

Given:

$$R(\tau)=16+\frac{9}{1+6\tau^2}.$$

To find:

Mean and variance

Formula to be used:

Variance $\sigma^2 = E\left[X^2(t)\right]-\left(E\left[X(t)\right]\right)^2$

$$\left(E\left[X(t)\right]\right)^2 = \lim_{|\tau|\to\infty} R_{xx}(\tau)$$

$$= \lim_{|\tau|\to\infty} 16+\frac{9}{1+6\tau^2}=16.$$

$$\therefore\ E\left[X(t)\right]=4.$$

$$\left(E\left[X^2(t)\right]\right)=R_{xx}(0)=16+\frac{9}{1+0}=16+9=25$$

∴ Variance $\sigma^2 = E\left[X^2(t)\right]-\left(E\left[X(t)\right]\right)^2$

$$= 25-(4)^2=9.$$

Mean = 4 and Variance = 9.

3. Let us calculate the mean and variance of a stationary random process whose auto correlation function is given by $R_{xx}(\tau)=18+\frac{2}{6+\tau^2}$.

Solution:

Given:

$$R_{xx}(\tau)=18+\frac{2}{6+\tau^2}$$

To find:

Mean and variance

Formula to be used:

Variance $\sigma^2 = E\left[X^2(t)\right] - \left(E\left[X(t)\right]\right)^2$

Given $R_{xx}(\tau) = 18 + \frac{2}{6+\tau^2}$

$$\bar{X}^2 = \lim_{|\tau|\to\infty} R_{xx}(\tau)$$

$$= \lim_{|\tau|\to\infty}\left[18 + \frac{2}{6+\tau^2}\right]$$

$$= 18 + \lim_{|\tau|\to\infty}\frac{2}{6+\tau^2}$$

$$= 18 + \frac{2}{6+\infty}$$

$$= 18 + 0$$

$$= 18$$

$$\bar{X} = \sqrt{18}$$

$$E\left[X(t)\right] = \sqrt{18}$$

$$\text{Var}\{X(t)\} = E\left[X^2(t)\right] - \left\{E\left[X(t)\right]\right\}^2$$

We know that,

$$E\left[X^2(t)\right] = R_{XX}(0)$$

$$= 18 + \frac{2}{6+0} = \frac{55}{3}$$

4. Let us find the auto correlation function of the periodic time function $\left[x(t)\right]$ = A sin ωt.

Solution:

Given:

$$\left[x(t)\right] = A \sin \omega t$$

Formula to be used:

$$R_{xx}(\overline{c}) = E[x(t) \times (t+\overline{c})$$

$$= E\left[A \sin \omega t \,.\, A \sin(\omega t + \omega \overline{c})\right)$$

$$= \frac{A^2}{2} E[\cos(\omega \overline{c}) - \cos(2\omega t + t\overline{c})]$$

sin A sin B = 1/2[cos(A - B) - cos(A + B)]

$$= \frac{A^2}{2} E(\cos \omega \overline{c}) - \frac{A^2}{2} E(\cos(2\omega t + \omega \overline{c}))$$

$$R_{xx}(\overline{c}) = \frac{A^2}{2} \cos \omega \overline{c} - \frac{A^2}{2} E[\cos(2\omega t + t\overline{c})]$$

4.8.1 Cross-correlation

$$\rho_{XY}(t_1, t_2) = \frac{c_{XY}(t_1, t_2)}{\sqrt{V_{ar}(X(t_1)) \mathrm{Var}(X(t_2))}}$$

Let {X(t)} and {Y(t)} be two random. Then the cross correlation between them is also defined as,

$$R_{XY}(t, t+\tau) = E[X(t)\, Y(t+\tau)]$$

$$= R_{XY}(\tau)$$

If the processes {X(t)} and {Y(t)} are jointly wide-sense stationary, then $E[X(t) \,.\, Y(t-\tau)]$ is a function of τ, denoted by $R_{xy}(\tau)$. This function $R_{xy}(\tau)$ is called as cross correlation function of the processes X(t) and Y(t).

If $R_{xy}(\tau) = \mu_x \times \mu_y$, then the processes X(t) and Y(t) are independent.

Properties of cross correlation:

1. The cross correlation function $R_{xy}(\tau)$ is not an even function of r. But it satisfies the condition,

$$R_{xy}(\tau) = R_{xy}(-\tau)$$

$$R_{xy}(\tau) = E[X(t)Y(t+\tau)] = E[X(p-\tau)Y(P)]$$

$$= R_{yx}(-\tau)$$

$$R_{xy}(\tau) = R_{yx}(-\tau)$$

$$R_{XY}(\tau) = E[X(t)Y(t+\tau)] = E[X(P-\tau)Y(P)]$$
$$= R_{YX}(-\tau)$$
$$R_{XY}(\tau) = R_{YX}(-\tau)$$

The above equation represents that $R_{xy}(\tau)$ and $R_{yx}(\tau)$ are mirror images of each other as depicted in figure.

2. Let X(t) and Y(t) are two random processes with respective autocorrelation functions $R_{xx}(\tau)$ and $R_{yy}(\tau)$.

Then, $|R_{xy}(\tau) \leq \sqrt{(R_{xx}(0)R_{YY}(0))}$

(i.e.) Cross correlation is bounded by geometric mean of $R_{xx}(0)$ and $R_{yy}(0)$.

3. The cross-correlation is bounded by arithmetic average of $R_{xx}(0)$ and $R_{yy}(0)$.

$$|R_{XY}(\tau)| \leq \frac{R_{XX}(0) + R_{YY}(0)}{2}$$

4. If X(t) and Y(t) are differentiable random processes then,

$$R_{XY}(\tau) = E[\dot{X}(t)\dot{Y}(t+\tau)] = \frac{d}{d\tau}R_{XY}(\tau) \text{ and}$$

$$R_{XY}(\tau) = E[\dot{X}(t)\dot{Y}(t+\tau)] = -\frac{d^2}{d\tau^2}R_{XY}(\tau)$$

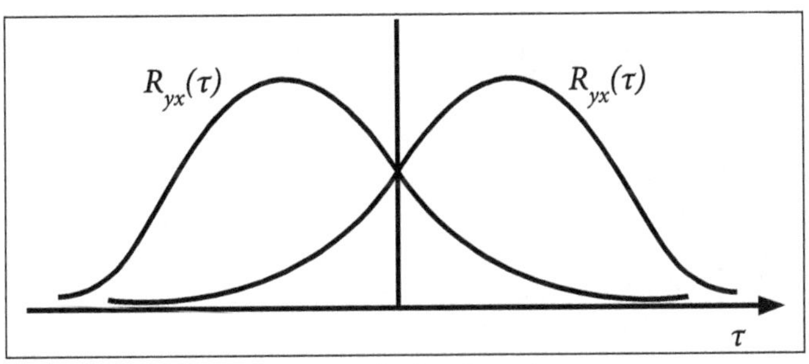

Cross correlation function.

5. If X(t) and Y(t) are uncorrelated random processes, then

$$R_{XY}(\tau) = E[X(t] \; E[Y(T+\tau)]$$

$$=E[X(t] E[Y(T)]$$

6. If X(t) and Y(t) are orthogonal random processes, then $R_{xy}(t, t + \tau) = 0$.

Problems:

1. Two random processes X(t) and Y(t) are given by $X(t) = A\cos(\omega t + \theta)$ and $Y(t) = A\sin(\omega t + \theta)$ where A and ω are constants and θ is a uniform random variable over $(0, 2\pi)$. For this case, let us determine the cross correlation function of X(t) and Y(t).

Solution:

Given:

$$X(t) = A\cos(\omega t + \theta)$$

$$Y(t) = A\sin(\omega t + \theta)$$

To find:

Cross correlation function of X(t) and Y(t)

Formula to be used:

$$R_{XY}(\tau) = R_{XY}(t, t + \tau)$$

$$R_{XY}(t, t + \tau) = E[X(t).Y(t + \tau)]$$

$$= E[A\cos(\omega t + \theta).A\sin(\omega(t + \tau) + \theta)]$$

$$= A^2 E[\sin\{\omega(t + \tau) + \theta\}\ \cos(\omega t + \theta)]$$

$$f(0) = \frac{1}{2\pi},\quad 0 \le \theta \le 2\pi$$

Now $E[\sin\{\omega(t + \tau) + \theta\}\cos(\omega t + \theta)]$

$$= \int_{-\infty}^{\infty} \sin(\omega t + \omega\tau + \theta).\cos(wt + \theta) f(\theta) d\theta$$

$$= \int_{0}^{2\pi} \sin(\omega t + \omega^{t+\theta}).\cos(\omega t + \theta)\left(\frac{1}{2\pi}\right) d\theta$$

$$=\frac{1}{2\pi}\int_{0}^{2\pi}\sin(\omega t+\omega\tau+\theta)\cos(\omega t+\theta)d\theta$$

$$=\frac{1}{2\pi}\int_{0}^{2\pi}\frac{1}{2}\{\sin(\omega t+\omega\tau+\theta+\omega t+\theta)+\sin[\omega t+\omega\tau+\theta-\omega t-\theta]\}d\theta$$

$$=\frac{1}{2\pi}\int_{0}^{2\pi}\frac{\sin[2\omega t+\omega\tau+2\theta]+\sin(\omega\tau)}{2}d\theta$$

Since θ is a uniformly distributed random variable we have,

$$=\frac{1}{4\pi}\left(-\frac{\cos(2\omega t+\omega\tau+2\theta)}{2}+\sin\,\omega\tau(\theta)\right)_{0}^{2\pi}$$

$$=\frac{1}{4\pi}\left[-\frac{\cos(2\omega t+\omega\tau)}{2}+\frac{\cos(2\omega t+\omega\tau+0)}{2}+\sin\,\omega\tau(2\pi-0)\right]$$

$$=\frac{1}{4\pi}\left[-\frac{\cos(2\omega t+\omega\tau)}{2}+\frac{\cos(2\omega t+\omega\tau)}{2}+2\pi\sin\omega\tau\right]$$

$$=\frac{1}{4\pi}[0+2\pi\sin\omega\tau]$$

$$=\frac{1}{2}\sin\omega\tau$$

$$R_{XY}(t,t_t)=\frac{A^2}{2}\sin\,\omega\tau$$

2. If the cross power spectral density of X(t) and Y(t) is,

$$S_{XY}(\omega)=\begin{cases}\alpha+\dfrac{ib\omega}{\alpha}; & -\alpha<\omega<a,\ >0\\ 0 & \text{otherwise}\end{cases}$$

where a and b are constants. Let us calculate the cross correlation function.

Solution:

Given:

$$S_{XY}(\omega)=\begin{cases}\alpha+\dfrac{ib\omega}{\alpha}; & -\alpha<\omega<a,\ >0\\ 0 & \text{otherwise}\end{cases}$$

To find:

Cross correlation function

Formula to be used:

The cross correlation function is,

$$R_{XY}(\tau)=\frac{1}{2\pi}\int_{-\infty}^{\infty}S_{XY}(\omega)e^{i\omega t}d\omega$$

$$=\frac{1}{2\pi}\int_{-a}^{a}\left(a+\frac{ibw}{a}\right)e^{i\omega\tau}d\omega$$

$$=\frac{1}{2\pi}\left[\left(a+\frac{ibw}{a}\right)\left(\frac{e^{i\omega\tau}}{i\tau}\right)-\left(\frac{ib}{a}\right)\left(\frac{e^{i\omega\tau}}{i^2\tau^2}\right)\right]_{-a}^{a}$$

$$=\frac{1}{2\pi}\left[(a+ib)\left(\frac{e^{i\omega\tau}}{i\tau}\right)-\frac{ib}{a\tau^2}e^{ia\tau}-(a-ib)\left(\frac{e-i\omega\tau}{i\tau}\right)+\frac{ib}{a\tau^2}e^{-ia\tau}\right]$$

$$=\frac{1}{2\pi}\left[\frac{a}{i\tau}\left(e^{ia\tau}-e^{-ia\tau}\right)+\frac{b}{\tau}\left(e^{ia\tau}+e^{-ia\tau}\right)+\frac{ib}{a\tau^2}\left(e^{ia\tau}+e^{-ia\tau}\right)\right]$$

$$R_{XY}(\tau)=\frac{1}{2\pi}\left[\frac{a}{\tau}(2\sin a\tau)+\frac{b}{\tau}(2\cos a\tau)-\frac{b}{a\tau^2}(2\sin a\tau)\right]$$

$$=\frac{1}{\pi}\left[\frac{a\sin a\tau}{\tau}+\frac{b\cos a\tau}{\tau}-\frac{b\sin a\tau}{a\tau^2}\right]$$

3. If the cross-power spectrum of real random processes {X(t)} and {Y(t)} is given by $S_{XY}(\omega)=\begin{cases}a+bj\omega, & \text{for } |\omega|<1\\ 0, & \text{elsewhere}\end{cases}$, then the cross correlation function is calculated as follows.

Solution:

Given:

$$S_{XY}(\omega)=\begin{cases}a+bj\omega, & \text{for } |\omega|<1\\ 0, & \text{elsewhere}\end{cases}$$

To find:

Cross correlation function

Formula to be used:

The cross-correlation function is given by,

$$R_{XY}(\tau) = \frac{1}{2\pi}\int_{-\infty}^{\infty} S_{XY}(\omega)e^{i\omega\tau}d\omega$$

$$= \frac{1}{2\pi}\int_{-1}^{1}(a+jb\omega)e^{j\omega\tau}\,d\omega$$

$$= \frac{1}{2\pi}\left[(a+jb\omega)\left(\frac{e^{j\omega\tau}}{j\tau}\right)-(jb)\left(\frac{e^{j\omega\tau}}{-\tau^2}\right)\right]_{-1}^{1}$$

$$= \frac{1}{2\pi}\left[\left\{(a+jb)\left(\frac{e^{j\tau}}{j\tau}\right)+(jb)\frac{e^{j\tau}}{\tau^2}\right\}-\left\{(a-jb)\frac{e-j\tau}{j\tau}+\frac{jb\ e^{-j\tau}}{\tau^2}\right\}\right]$$

$$= \frac{1}{2\pi}\left[\frac{a}{j\tau}\left(e^{j\tau}-e^{-j\tau}\right)+\frac{b}{\tau}\left(e^{j\tau}+e^{-j\tau}\right)-\frac{jb}{\tau^2}\left(e^{j\tau}-e^{-j\tau}\right)\right]$$

$$= \frac{1}{2\pi}\left[\frac{a}{j\tau}2j\sin\tau+\frac{b}{\tau}2\cos\tau-\frac{jb}{\tau^2}2j\sin\tau\right]$$

$$= \frac{1}{\pi}\left[\frac{a}{\tau}\sin\tau+\frac{6}{\tau}\cos\tau-\frac{b}{\tau^2}\sin\tau\right]$$

$$\therefore R_{XY}(\tau) = \frac{1}{\pi r^2}\left[(a\tau-b)\sin\tau+b\tau\cos\tau\right]$$

4.8.2 Power Spectral Density

The power spectrum is transformed when the process is filtered by a linear time invariant filter. Let Y(t) be the output of a linear time invariant filter with frequency response H(f) and X(t) is the input signal, then the power spectral density of the output process Y(t) is given as,

$$S_Y(f) = S_X(f)|H(f)|^2$$

If {X(t)} is a stationary process with autocorrelation function $R_{XX}(t)$, then the Fourier transform of $R_{XX}(t)$ is called as the power spectral density function of {X(t)} or simply spectral density of X(t) and is given as,

$S_{XX}(\omega) =$ Fourier Transform of $R_{XX}(\tau)$

$$= \int_{-\infty}^{\infty} R_{XX}(\tau) e^{-i\omega\tau} d\tau$$

Thus,

$$S_{XX}(f) = \int_{-\infty}^{\infty} R_{XX}(\tau) e^{-i2\pi f\tau} d\tau$$

Power Spectral Density and Autocorrelation

The energy spectral density and autocorrelation function of energy signals are important tools for the characterization of energy signals. Power signals are infinite in time. It remains finite as time t approaches infinity and so their energy is infinite.

Power signals are very important in telecommunications, since they include,

- Periodic signals
- Channel white noise

Most transmitted signals are either analog or digital signals.

Therefore for power signals, it is desirable to have a counter part of the energy spectral density and autocorrelation function of energy signals. They are called as power spectral density and autocorrelation function of power signals.

In time domain, we define average power as,

$$P_x = \lim_{T_0 \to +\infty} \frac{1}{2T_0} \int_{-T_0}^{+T_0} |x(t)|^2 dt$$

Which is finite but non-zero.

In frequency domain, the description of power might not be possible because the signal is infinite and may not have Fourier transform.

The simplest way to define the PSD is by assuming that our infinite duration signal is the limit for a proper finite duration signal, (i.e.),

$$x(t) = \lim_{T_0 \to +\infty} x_{T_0}(t)$$

With $x_{T_0}(t)$ given by,

$$x_{T_0}(t)=\begin{cases} x(t), & |t|<T_0 \\ 0, & \text{otherwise} \end{cases}$$

So, the above equation for average power becomes,

$$P_x=\lim_{T_0\to+\infty}\frac{1}{2T_0}\int_{-\infty}^{+\infty}\left|x_{T_0}(t)\right|^2 dt$$

From the Rayleigh theorem,

$$\int_{-\infty}^{+\infty}\left|x_{T_0}(t)\right|^2 dt=\int_{-\infty}^{+\infty}\left|X_{T_0}(f)\right|^2 df$$

With,

$X_{T_0}(f)=\int_{-\infty}^{+\infty}x_{T_0}(t)\exp(-j2\pi ft)dt$ denoting the Fourier transform of $x_{T_0}(t)$. Note that $X_{T_0}(t)$ is finite and has Fourier transform. Now, the average power can be written in terms of frequency as,

$$P_x=\lim_{T_0\to+\infty}\frac{1}{2T_0}\int_{-\infty}^{+\infty}\left|X_{T_0}(f)\right|^2 df$$

The convergence of the integral will allow us to include the limit inside the integrand and therefore, the PSD of x(t) is given as,

$$S_x(f)=\lim_{T_0\to+\infty}\frac{1}{2T_0}\left|X_{T_0}(f)\right|^2$$

In general, the energy of the signal $x_{T_0}(t)$ associated to a power signal grows approximately linearly with T_0. Therefore, the limit associated to PSD exists and is finite.

However, this approach for computing the PSD has two difficulties:

- For most power signals it is difficult to obtain $X_{T_0}(f)$ and the limit associated to PSD. This is especially true for random power signals.
- The limit can be infinite for some power signals. For instance, if $x_{T_0}(t)=A$ then $X_{T_0}(f)=AT_0\,\text{sinc}(fT_0)$, which means that $S_x(0)=+\infty$.

To overcome these difficulties, we can define an autocorrelation function of power signals and then relate it with PSD. Essentially, we can define the PSD of x(t) as,

$$S_x(f) = \int_{-\infty}^{+\infty} R_x(\tau)\exp(-j2\pi ft)dt$$

Where, $R_x(\tau)$ denotes the autocorrelation of x(t) and is defined as,

$$R_x(\tau) = \lim_{T_0 \to +\infty} \frac{1}{2T_0} \int_{-T_0}^{T_0} x(t)x^*(t-\tau)dt$$

This means that the PSD of x(t) is the Fourier transform of its autocorrelation.

The average power of x(t) is given by,

$$P_x = \int_{-\infty}^{+\infty} S_x(f)df$$

From the autocorrelation definition,

$$P_x = R_x(0)$$

The average power, PSD and autocorrelation have the following properties:

1. $P_x = \int_{-\infty}^{+\infty} S_x(f)df$
2. $P_x = R_x(0)$
3. The power of the signal associated to the band $[f_1, f_2]$ is given by,

   $$\int_{f_1}^{f_2} S_x(f)df$$

4. $S_x(f) = \mathcal{F}\{R_x(\tau)\} = \int_{-\infty}^{+\infty} R_x(\tau)\exp(-j2\pi ft)dt$
5. $R_x(\tau) = \mathcal{F}^{-1}\{S_x(f)\} = \int_{-\infty}^{+\infty} S_x(f)\exp(j2\pi ft)dt$
6. $|R_x(\tau)| \leq R_x(0)$
7. $S_x(f) \geq 0$
8. $R_x(-\tau) = R_x^*(\tau)$
9. For real signal, $R_x(-\tau) = R_x(\tau)$ and $S_x(-f) = S_x(f)$

10. If x(t) is submitted to a filter with impulse response h(t) and frequency response H(f), then the resulting signal y(t) = x(t)*h(t) has PSD,

$$S_y(f) = S_x(f)|H(f)|^2$$

Its autocorrelation is expressed as,

$$R_y(\tau) = R_x(\tau) * h(t) * h^*(-t)$$

Properties of Power Density Spectrum

The properties of power density spectrum $S_{XX}(\omega)$ for a WSS random process X(t) are given as,

1. $S_{XX}(\omega) \geq 0$

Proof:

The expected value of a non-negative function $E\left[\left|X_T(\omega)\right|^2\right]$ is always non-negative.

Therefore, $S_{XX}(\omega) \geq 0$ hence proved.

2. The power spectral density at zero frequency is equal to the area under the curve of the autocorrelation $R_{XX}(\tau)$, (i.e.) $S_{XX}(0) = \int_{-\infty}^{\infty} R_{XX}(\tau)_{d\tau}$

Proof:

We know that, $S_{XX}(\omega) = \int_{-\infty}^{\infty} R_{XX}(\tau) e^{-j\omega\tau} d\tau$ at $\omega = 0$,

$$S_{XX}(0) = \int_{-\infty}^{\infty} R_{XX}(\tau)\, d\tau$$

Hence proved.

3. The power density spectrum of a real process X(t) is an even function, (i.e.) $S_{XX}(-\omega) = S_{XX}(\omega)$.

Proof:

Consider a WSS real process X(t), then,

$$S_{XX}(\omega) = \int_{-\infty}^{\infty} R_{XX}(\tau) e^{-j\omega\tau} d\tau \text{ also } S_{XX}(-\omega) = \int_{-\infty}^{\infty} R_{XX}(\tau) e^{j\omega\tau} d\tau$$

Substitute τ = -τ then,

$$S_{XX}(-\omega) = \int_{-\infty}^{\infty} R_{XX}(-\tau) e^{-j\omega\tau} d\tau$$

Since X(t) is real, we know that, $R_{XX}(-\tau) = R_{XX}(\tau)$.

Therefore,

$$S_{XX}(-\omega) = \int_{-\infty}^{\infty} R_{XX}(\tau) e^{j\omega\tau} d\tau$$

$S_{XX}(-\omega) = S_{XX}(\omega)$ Hence proved.

4. $S_{XX}(\omega)$ is always a real function.

Proof:

We know that, $S_{XX}(\omega) = \lim_{T\to\infty} \dfrac{E\left[\left|X_{T(\omega)}\right|^2\right]}{2T}$

Since the function $\left|\left|X_{T(\omega)}\right|^2\right|$ is a real function, $S_{XX}(\omega)$ is always a real function hence proved.

5. If $S_{XX}(\omega)$ is a PSD of a WSS random process X(t), then,

$\frac{1}{2\pi}\int_{-\infty}^{\infty} S_{XX}(\omega)_{d\omega} = A\{E[X^2(t)]\} = R_{XX}(0)$ or the time average of the mean square value of a WSS random process equals the area under the curve of the power spectral density.

Proof:

We know that, $R_{XX}(\tau) = A\{E[X(t+\tau)X(t)]\}$

$$= \frac{1}{2\pi}\int_{-\infty}^{\infty} S_{XX}(\omega)_{d\omega} \text{ at } \tau = 0,$$

$R_{XX}(0) = A\{E[X^2(t)]\} = \frac{1}{2\pi}\int_{-\infty}^{\infty} S_{XX}(\omega)_{d\omega}$ = Area under the curve of the power spectral density.

Hence proved.

6. If X(t) is a WSS random process with PSD $S_{\dot{X}\dot{X}}(\ )$, then the PSD of the derivative of X(t) is equal to ω^2 times the PSD $S_{XX}(\omega)$.

That is, $S_{\dot{X}\dot{X}}(\omega) = \omega^2 S_{XX}(\omega)$.

Proof:

$$X_T(\omega) = \int_{-T}^{T} X(t) e^{-j\omega t} dt$$

We know that $S_{XX}(\omega) = \lim_{T\to\infty} \frac{E\left[\left|X_T(\omega)\right|^2\right]}{2T}$ and

$$= \dot{X}_T(\omega) = \frac{d\,X_T(\omega)}{dt} = \frac{d}{dt}\int_{-T}^{T} X(t)e^{-j\omega t}dt$$

$$= \int_{-T}^{T} X(t)\frac{d}{dt}e^{-j\omega t}dt$$

$$= \int_{-T}^{T} X(t)(-j\omega)e^{-j\omega t}dt$$

$$= (-j\omega)\int_{-T}^{T} X(t)e^{-j\omega t}dt$$

Therefore, $S_{\dot{X}\dot{X}}(\omega) = \lim_{T\to\infty} \frac{E\left[\left|\dot{X}_T(\omega)\right|^2\right]}{2T}$

$$= \lim_{T\to\infty} \frac{E\left[\left|(-j\omega)X_T(\omega)\right|^2\right]}{2T}$$

$$= \lim_{T\to\infty} \frac{\omega^2 E\left[\left|(-j\omega)X_T(\omega)\right|^2\right]}{2T}$$

$$S_{\dot{X}\dot{X}}(\omega) = \omega^2 \lim_{T\to\infty} \frac{E\left[\left|\dot{X}_T(\omega)\right|^2\right]}{2T}$$

Therefore, $S_{\dot{X}\dot{X}}(\omega) = \omega^2 S_{XX}(\omega)$ hence proved.

Problems

1. If $R(\tau) = e^{-2\lambda|\tau|}$ is the autocorrelation function of a random process {X(t)}, then let us determine the spectral density of {X(t)}.

Solution:

Given:

$$R(\tau) = e^{-2\lambda|\tau|}$$

$$S(\omega) = \int_{-\infty}^{\infty} E(\tau)e^{-i\omega\tau}d\tau$$

Formula to be used:

$$S(\omega) = \int_{-\infty}^{\infty} R(\tau)e^{-i\omega\tau}d\tau$$

$$= \int_{-\infty}^{\infty} e^{-2\lambda|\tau|}(\cos\omega\tau - i\sin\omega\tau)d\tau$$

$$= 2\int_{0}^{\infty} e^{-2\lambda\tau}\cos\omega\tau d\tau$$

$$= \left[\frac{2e^{-2\lambda\tau}}{4\lambda^2+\omega^2}(-2\lambda\cos\omega\tau + \omega\sin\omega\tau)\right]_0^{\infty}$$

$$S(\omega) = \frac{4\lambda}{4\lambda^2+\omega^2}$$

2. A wide sense stationary noise process N(t) has an autocorrelation function $R_{NN}(\tau) = Pe^{-3|\tau|}$, where P is a constant. Let us calculate its power spectrum.

Solution:

Given:

$$R_{NN}(\tau) = Pe^{-3|\tau|}$$

$$S_{XX}(\omega) = \int_{-\infty}^{\infty} e^{-j\omega\tau}R_{XX}(\tau)d\tau$$

Formula to be used:

$$S_{NN}(\omega) = P\int_{-\infty}^{\infty} e^{-3|\tau|}(\cos\ \omega\tau - j\ \sin\ \omega\tau)d\tau$$

$$= P\int_{-\infty}^{\infty} e^{-3|\tau|}\cos\ \omega\tau d\tau - Pj\int_{-\infty}^{\infty} e - 3|\tau|\ \sin\ \omega\tau d\tau$$

$$= 2P\int_{0}^{\infty} e^{-3\tau}\cos\ \omega\tau\ d\tau - o \text{ (By properties of defining integral)}$$

$$= 2P\left[\frac{e^{-3\tau}}{9+\omega^2}(-3\cos\omega\tau + \omega\sin\omega\tau)\right]_0^{\infty}$$

$$S_{NN}(\omega) = \frac{6p}{9+\omega^2}$$

3. Let us determine the power spectral density of a random signal with autocorrelation function $e^{-\lambda|\tau|}$.

Solution:

Given:

$$e^{-\lambda|\tau|}$$

$$S(\omega) = \int_{-\infty}^{\infty} R(\tau)e^{-i\omega\tau}d\tau$$

Formula to be used:

$$S(\omega)=\int_{-\infty}^{\infty} R(\tau)e^{-i\omega\tau}d\tau$$

$$=\int_{-\infty}^{\infty} e^{-\lambda|\tau|}(\cos\omega\tau-i\sin\omega\tau)d\tau$$

$$=2\int_{0}^{\infty} e^{-\lambda|\tau|}\cos\omega\tau\, d\tau$$

$$=2\left[\frac{e^{-\lambda\tau}}{\lambda^2+\omega^2}(-\lambda\cos\omega\tau+\omega\sin\omega\tau)\right]_0^{\infty}$$

$$=2\left[0-\frac{1}{\lambda^2+\omega^2}(-\lambda)\right]=\frac{2\lambda}{\lambda^2+\omega^2}$$

4.9 Noise

Noise Sources and Types

Noise is a random, undesirable electrical energy that enters the communication system through the communicating medium and interferes with the transmitted message. However, some noise is also produced in the receiver.

Noise may be defined as any unwanted form of energy which tends to interfere with proper reception and reproduction of wanted signal.

Noise may be classified into two types. They are:

- Internal Noise.
- External Noise.

Internal noise can be easily evaluated mathematically and can be reduced to a great extent by proper design. External noise on the other hand cannot be reduced, except by changing the location of the receiver or the entire system.

Internal Noise in Communication

Internal noise is the noise which gets generated within the receiver or communication system. Internal noise may be put into the following four categories:

- Shot noise.
- Thermal noise or white noise or Johnson noise.

- Miscellaneous internal noise.
- Transit time noise.

Shot Noise: The most common type of noise is referred to as shot noise, which is produced by the random arrival of electrons or holes at the output element, at the plate in a tube or at the collector or drain in the transistor.

Shot noise is also produced by the random movement of electrons or holes across a PN junction. Even though current flow is established by external bias voltage, there will be some random movement of electrons or holes due to the discontinuities in the device.

An example of such discontinuity is the contact between the copper lead and the semiconductor materials. The interface between the two creates discontinuity that causes random movement of the current carriers.

Thermal Noise: Conductors contain a large number of free electrons and ions that are strongly bounded by molecular forces. The ions vibrate randomly about their normal positions; however, this vibration is a function of temperature. Continuous collisions between the electrons and vibrating ions take place. Hence, there is a continuous transfer of energy between the ions and electrons. This is the source of resistance in a conductor.

The movement of free electrons constitutes a current which is purely random in nature and over a long time, averages zero. There is a random motion of electrons which gives rise to noise voltage called as thermal noise. Hence, the noise generated in any resistance due to the random motion of electrons is called as thermal noise or white noise or Johnson noise.

The analysis of thermal noise is based on the Kinetic theory. It shows that, the temperature of particles is a way of expressing its internal kinetic energy. Therefore, temperature of a body can be said to be equivalent to the statistical RMS value of the velocity of motion of particles in the body.

At -273°C, the kinetic energy of the body becomes zero. Hence, we can relate the noise power generated by the resistor to be proportional to its absolute temperature. Noise power is also proportional to the bandwidth over which noise is measured.

$$P_n \propto TB$$

$$P_n = KTB \qquad \text{...(1)}$$

Where,

P_n = Maximum noise power output of the resistor.

K = Boltzmann's constant = 1.38 x 10^{-23} joules/Kelvin.

T = Absolute temperature.

B = Bandwidth over which noise is measured.

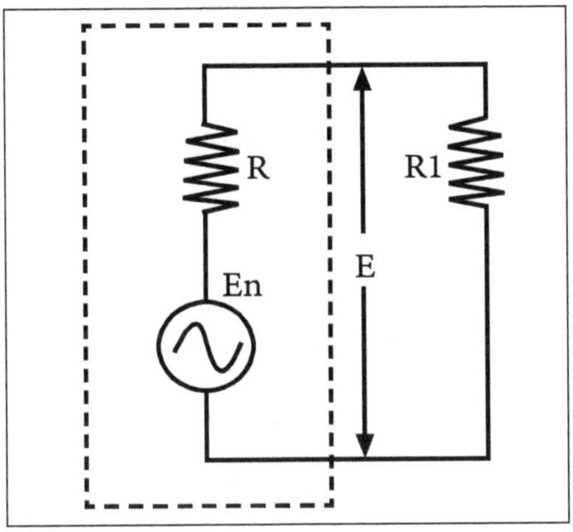

$$P_n = \frac{E^2}{R_1} = \frac{E^2}{R} = \frac{\left(\frac{E_n}{2}\right)^2}{R} = \frac{E_n^2}{4R}$$

Therefore, from above figure if $R = R_2$, $E_n = E/2$.

$$E_{n2} = 4RP_n$$

$$E_{n2} = 4RKTB$$

$$E_n = \sqrt{4KTRB} \qquad ...(2)$$

But, when the frequency of operation is high and the signal being processed is the magnitude of transit time, then problem can occur.

Miscellaneous Internal Noise: Flicker noise or modulation noise is the one appearing in transistors operating at low audio frequencies. Flicker noise is proportional to the emitter current and junction temperature. However, it is inversely proportional to frequency. Hence, it may be neglected at frequencies above 500 Hz and therefore, possess no serious problems.

Transit Time Noise: Another kind of noise that occurs in the transistor is termed as transit time noise. Transit time is the duration of time that it takes for the current carrier such as a hole or current to move from input to the output.

The devices are very tiny, so the distances involved are minimal. Yet the time it takes for the current carriers to move even short distance is finite.

At low frequencies, this time is negligible, but when the frequency of operation is high and the signal being processed is the magnitude of the transit time, then problem can occur. Then transit time shows up as a kind of random noise within the device and this is directly proportional to the frequency of operation.

External Noises

External noise may be classified into the following three types:

- Extraterrestrial noise.
- Atmospheric noise.
- Man-made noises or industrial noise.

Atmospheric Noise: Atmospheric noise is caused by the lighting discharges in the thunderstorms and other natural electrical disturbances occurring in the atmosphere. These electrical impulses are random in nature. Thus, the energy is spread over the complete frequency spectrum used for radio communication.

Atmospheric noise consists of spurious radio signals with components spread over a wide range of frequencies. These spurious radio waves constitute the noise; get propagated over the earth in the same fashion as the desired radio waves of the same frequency.

Accordingly at a given receiving point, the receiving antenna picks up not only the signal, but also the static energy from all the thunderstorms, local or remote.

Extraterrestrial Noise:

- Cosmic noise.
- Solar noise.

Cosmic Noise: Sun is known as the distant star and has high temperatures. Therefore, these stars radiate noise in the same way as sun. The noise received from these distant stars is thermal noise and these get distributed uniformly over the entire sky. We also receive the noise from the center of our own galaxy, other distant galaxies and from other virtual point sources such as quasars and pulsars.

Solar Noise: This is the electrical noise emanating from the sun. Under quite condition, there is a steady radiation of noise from the sun. This results, because sun is a large body which has very high temperature and radiates electrical energy in the form of noise over a wide frequency spectrum including the spectrum used for radio communication. The intensity interference make many frequencies unusable for communications. During other years, the noise is at a minimum level.

Man-made Noise or Industrial Noise: The electrical noise is produced by sources such as electrical motors and switch gears, automobiles and aircraft ignition, fluorescent lights, leakage from high voltage lines and numerous other heavy electrical machines. Such noises are produced by the arc discharge taking place during the operation of these machines. Such man-made noise is intensive in industrial and densely populated areas.

Man-made noise in such areas far exceeds all other sources of noise in the frequency range extending from 1 MHz to 600 MHz.

Noise Removal and Signal Compensation

Removing noise from a signal is the most frequent application for signal processing. The distinction between noise and desired signal is a heuristic judgment that must be determined by the user. Often the distinction is obvious, for example, 60 Hz interference in an EEG recording or temperature related drift in the reading of a pH electrode. In other situations the user depends on experience and indirect knowledge of the signal and its source to decide whether or not certain features of a signal should be considered as noise.

For example, since the frequency components of the EEG are principally below 25 Hz, one would suspect that prominent signals in the EEG at much higher frequencies have another source such as the electromyogram activity of neck, jaw or extra ocular muscles. Once the distinction between the desired signal and noise has been resolved, then removal of the noise with the least possible distortion of the desired signal is the goal.

One may specify the frequency response of a linear filter that can affect the separation of the frequency components of the noise from those of the desired signal. However, in many cases, the necessary frequency response may not be similar to that of the simple first and second order filters. In other cases, the separation between the pass band and stop band of a simple filter may be inadequate, because either the stop band gain is not small enough or the frequency components of signal and noise are too close together relative to the width of the transition between the pass band and stop band.

In these situations, it is necessary to design a better filter, because biomedical applications often involve the acquisition of continuous time signals by digital sampling, one has the choice of filtering either before sampling using a continuous time filter or after sampling using a discrete time filter. Digital filters are more versatile and more convenient to modify for different applications.

However, analog filters also have certain advantages and there are pertinent reasons to employ them. First, to prevent aliasing one should always filter a CT signal with an analog filter before sampling it. Second, even if it is incomplete, any noise removal that one can effect with an analog filter before sampling reduces the requirements on any subsequent digital filter. Finally, the properties of some classes of analog filters are well studied and these filters provide a foundation for designing equivalent digital filters, because much filtering of biomedical signals is accomplished digitally.

Many newer biomedical instruments that acquire analog signals process these signals by digitizing them and often include built-in digital filters, because digital filters are designed using Z-transforms.

Permissions

All chapters in this book are published with permission under the Creative Commons Attribution Share Alike License or equivalent. Every chapter published in this book has been scrutinized by our experts. Their significance has been extensively debated. The topics covered herein carry significant information for a comprehensive understanding. They may even be implemented as practical applications or may be referred to as a beginning point for further studies.

We would like to thank the editorial team for lending their expertise to make the book truly unique. They have played a crucial role in the development of this book. Without their invaluable contributions this book wouldn't have been possible. They have made vital efforts to compile up to date information on the varied aspects of this subject to make this book a valuable addition to the collection of many professionals and students.

This book was conceptualized with the vision of imparting up-to-date and integrated information in this field. To ensure the same, a matchless editorial board was set up. Every individual on the board went through rigorous rounds of assessment to prove their worth. After which they invested a large part of their time researching and compiling the most relevant data for our readers.

The editorial board has been involved in producing this book since its inception. They have spent rigorous hours researching and exploring the diverse topics which have resulted in the successful publishing of this book. They have passed on their knowledge of decades through this book. To expedite this challenging task, the publisher supported the team at every step. A small team of assistant editors was also appointed to further simplify the editing procedure and attain best results for the readers.

Apart from the editorial board, the designing team has also invested a significant amount of their time in understanding the subject and creating the most relevant covers. They scrutinized every image to scout for the most suitable representation of the subject and create an appropriate cover for the book.

The publishing team has been an ardent support to the editorial, designing and production team. Their endless efforts to recruit the best for this project, has resulted in the accomplishment of this book. They are a veteran in the field of academics and their pool of knowledge is as vast as their experience in printing. Their expertise and guidance has proved useful at every step. Their uncompromising quality standards have made this book an exceptional effort. Their encouragement from time to time has been an inspiration for everyone.

The publisher and the editorial board hope that this book will prove to be a valuable piece of knowledge for students, practitioners and scholars across the globe.

Index

www.ingramcontent.com/pod-product-compliance
Lightning Source LLC
LaVergne TN
LVHW080639210926
848485LV00003B/295